The Maiden Voyage of Petrus van Stijn

Michael Charles Tobias

The Maiden Voyage of Petrus van Stijn

A Novel

 Springer

Michael Charles Tobias
President, Dancing Star Foundation
Los Angeles, CA, USA

ISBN 978-3-030-97685-9   ISBN 978-3-030-97683-5   (eBook)
https://doi.org/10.1007/978-3-030-97683-5

"Ross Sea, Antarctica," © By Craig Potton, Courtesy of Craig Potton

This Springer imprint is published by the registered company Springer Nature Switzerland AG
The registered company address is: Gewerbestrasse 11, 6330 Cham, Switzerland

This little story is dedicated to Jane Gray Morrison, the love of my life.

Foreword

Michael Tobias is one of our most profound and prolific writers. He has written a new book.

There are many of us who look to the future with troubling concern. The ecological tipping points may have been reached, or shortly *will* be reached. The coming years are going to be very disturbing for all of us.

It is easy to lose faith. To have a dark view of humanity.

But when a writer of Michael's breadth and depth has something positive to say, we should all run—not walk—to the nearest pages of that book and start reading at once.

Tobias' novel will give you hope. Without hope, where do we begin?

What a literate, imaginative, beautiful writer Michael Tobias has been throughout a long career!

The Maiden Voyage of Petrus van Stijn may well be Tobias' finest work.

Actor, Writer, Producer   William Shatner
November 6, 2021

Acknowledgments

I wish to thank my editor, Kenneth Teng, Publishing Editor in Life Sciences at Springer, for his belief in this work. And to Springer itself, for many wonderful collaborations on book projects over the years. I also want to acknowledge and thank deeply my wife and partner, Jane Gray Morrison, in all things. Our first sustained "date" was in Antarctica. Great thanks to the conservationist, writer, and photographer, Craig Potton, a longtime friend, for use of a few of his superb Antarctic photographs in the book. Such is the lure of the polar south that my late mother, Betty Mae Tobias, journeyed to Antarctica with my brother, Marc, when she was in her mid-90s. Finally, many thanks to a dear friend, William Shatner, for his Foreword, written not long after he *returned* to earth.

Contents

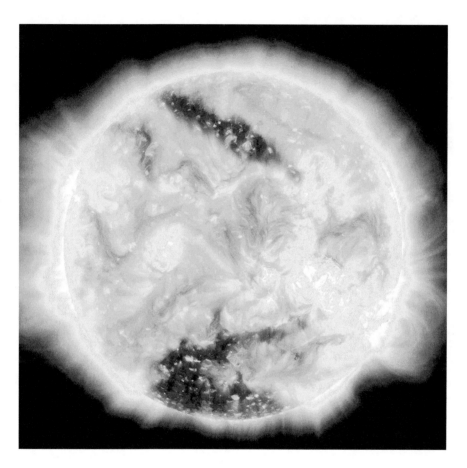

Two coronal holes on the Sun, © NASA's Solar Dynamics Observatory, March 16, 2015, NASA/SDO, Public Domain

Chapter 1
Doomsday Shelfs

Early August. Winter. 06:30. The signal had been pinging for days, emanating from the High Elevation Antarctic Terahertz telescope complex at Ridge A. Outside temperature was minus 157 °F. Urine, vodka, spit, there is very little that remains liquid in such weather. From ridges and dips over 1000mi away, the downward funneling katabatic winds roared inland, high-density multiples of ice tornadoes.

For 3 days incoming data just blipped erratically every 27–30 s in vague sputters, superposition qubits far beyond the quantum supremacy state, towards 10^{48}th. One needed merely to glance at the optics. Those remaining at the Belgian enclave used an ice-drilling periscope to punch through the enormous accumulation of snowfall that had continued to engulf Princess Elisabeth Station, or pes as she was known, for over 4 days. In such storm, standing on the roof, tied in, with a shovel, was simply a no-go.

Pes' weather analyst was fascinated and disturbed by the persistent anomalies, everyday a new surprise in store for the remaining survivors. Nothing like it had ever been observed. Not even if one looked back to the earliest records from NASA's old MODIS (spectroradiometer) on the Aqua satellite that had gone silent so many years before.

Remarkably, a mere 8 months prior to the present ice frenzies, temperatures had soared to more than 70°, killing additional scores of indigenous fauna. Such weather oscillations were by now the norm. But Hopf Bifurcations [1] were not. Abstruse, but real enough, they incited an entirely new, more ferocious level of ecological spasms altogether. With that chaos came a decision put forth by a small coterie of researchers calling for a more finely calibrated doomsday calendar. 2029 equaled (in *adj* -for adjusted) the year 2056, and so on. Hebrew, Vedic, Papal, Zodiacal, Islamic, Japanese, Australasian calendars had all historically linked to solstices, lunar appearance, and other natural phenomena. The new adj correlations were predicated upon such biological accelerations and de-accelerations as could be ascertained in the realms of genetic and metabolic interactions, photoperiodism, molecular dormancy, and newly divined circadian rhythms. For decades the

© The Author(s), under exclusive license to Springer Nature Switzerland AG 2022
M. C. Tobias, *The Maiden Voyage of Petrus van Stijn*,
https://doi.org/10.1007/978-3-030-97683-5_1

scientific community had understood the so-called SCN, or suprachiasmatic nuclei. In effect, SCN represented a kind of signaling from within the mammalian hypothalamus; the veritable timekeeper, with some ten-thousand neurons serving collectively to maintain a sense of biological coherence and ritual, day and night [2].

But for other species, the clock was ticking sideways, downwards, or in every sense awry. As untoward events necessitated a different chronology to account for the death, or transcendence of phytoplankton and countless larval forms, invertebrates, vertebrates. The new sequencing of days and months also incorporated the mutational fall-out from basal cell samples, as well as from body parts of plants and animals, as scrutinized in the few remaining laboratories.

While at first inconceivable, it appeared that novel genotypes were thriving with far greater statistical probability than behavior exhibited by most non-modified organisms.

The majority of people still lived in what was designated as the pre-adj (PA) world, unwilling to ascribe in their minds to the many scientific disciplines that had computed the syncretistic fallouts and mounting tolls of human overshoot. Of systems, like methane emissions, ammonia runoff from industrialized animal factories across Europe and Asia, runaway bacterial transductions, and solar fiascos that had collided atop the ruinous overshoots of the planet's most populous primate. A (adj) versus P(re)A had escalated into a time-bomb of stark and furious social division: those who denied the science versus others who lived by it.

Indeed, with so many live doomsday clocks, no one could actually keep time. Time no longer mattered. It was naked.

There was simply no universal solution in sight to impede the spate of instabilities. No discipline, neither physics, neurology nor biochemistry, offered a full-enough window on what was happening. To dwell on a world without "solutions" brought on only deeper depression in those who believed there was something that could be done. And no special theorem to explain anything anymore. But, at least, plenty of pills to curb the anxieties.

Like the shocking combination over the course of a year of 150 m of warm snow, the icy tempests, black fleas streaking 15-mi-wide rapidly moving floes of red slush (iron oxide), like bloody lava. Amid intermittent rain and lightning, and a rash of seismic activity that held captive this last hold-out of humanity. Its apparent certainty gave rise to an almost bemused frame of mind. There was little left to go wrong. And no re-assuring clarity emerging from this blurred and beleaguered community of scientists and technicians. They had been essentially trapped for decades in their own nearly closed ecological life support system.

The coastal ice shelf was poised to collapse with frozen faults, cracked and glistening, surrounding pes and extending all the way from the ocean. End of the world shelfs, undercut by warm waters, strong currents and high salinity. It had started half-century before across the continent at the West Antarctic ice sheet. These were known as "pinning points" throughout the sinking Aristotle Mountains, but particularly beneath Thwaites glacier, a heap of horrible data from the Ran submarine. And the combination of other, strongly ill-boding numbers commencing towards the end of the last century that yielded a window on extreme polar movement in rapid, flexuous directions, as well as actual gravity and earth axis shifts. These were deemed to have resulted not only from human-induced re-location of water sheets (over one hundred trillion metric tons—as of the 2050s), but the melting, or partial decay of every one of over 200,000 glaciers on the planet.

Since the pre-adj time the *lights went out*, most of the newly denominated *storm cloud* back-up records had been compromised or lost entirely. It had been theorized at CERN [3] that there was a peculiar interaction occurring between quarks and the Higgs boson which was engendering a spasmatic, intensively penetrating new sub-atomic particle. Inferences arising from such computational techniques as the lattice quantum chromodynamic calculations [4] had been stirring debate for over half-century. To some, the end particle, the first particle as many preferred to think of it, had signaled to the remaining few, an inalienable finale. It appeared to be too late for increments or compromises. A vast dying, with no community to grasp, or even attempt to rationalize the losses. For a few years, in the beginning, governments had scrambled to fix ecological taxes on everything. But the so-called sustainability credits, and the reformed banking systems to embrace had long ceased to be enforceable.

Critical tallies, like those from the German aerospace Gravity Recovery and Climate Experiment (GRACE) satellites, were dead. Researchers were left to extrapolate earth axis changes in absence of any new information. One scientist described it as the day Euclid, or Eratosthenes was born. Starting almost from the beginning, in other words. Scientists and mathematicians, poets and dramatists with no peerage, no audience, as if all were howling into the dark.

While ferocious blizzards in winter and summer were nothing new, ice torna-does, and certainly such aggravated cold, had never been documented. Despite the uncanny engineering and years of sturdy retrofitting at the station, a novel fear had enshrouded the remaining occupants, as the 50th anniversary of the construction of pes had come and long gone. Leakage in dozens of areas. Unaccountable glacial acoustics. And the new, previously unthinkable, sense that even the unshakable architectural dynamics of Belgium's key presence in Antarctica, a complex whose construction was the wonder of the continent, when it was first airlifted into place, was no longer immune to something very frightening.

In the aftermath of that strangely bizarre anniversary was left the odd niche of useless philosophizing. Thoughts arose, then burnt out in a world-isolated angst and everyday reminders of the degree of human fragility. These syndromes of a mental plight corresponded with absolute negative temperatures on a continent that was breaking up in its apparently end of days. The weary occupants' imaginations roamed across a vestigial, caudal memory of the last Ice Ages. But that was hardly the comforting yarn.

As if the isolation were not suffocating enough. Even memories had been dulled between a living fiction and the myriad of whiteouts comprising the past, each one segueing into the next.

Then the station witnessed a second suicide within a week. Both scientists were buried and briefly mourned. Strange tears froze to one's face. Back indoors, slam-ming the steel doors hard against the blowing exterior inferno furthered the compulsion to get under one's covers, curl up, eyes closed, and just disappear. Dozens had done so over the last decade or so. Magical, instant farewells.

References

1. Named after three 20th century mathematicians and physicists -Eberhard Hopf, Aleksandr Andronov and Henri Poincaré – who quantified how a system could shift suddenly from stability to instability
2. See "Neuroanatomy, Nucleus Suprachiasmatic," by Melinda A. Ma and Elizabeth H. Morrison, [Updated 2021 Jul 31]. In: StatPearls [Internet]. Treasure Island (FL): StatPearls Publishing; 2021 Jan-. Available from: https://www.ncbi.nlm.nih.gov/books/NBK546664/; See also, "Toward a Molecular Biological Calendar?" by Michael H. Hastings and Brian K. Follett, Journal Of Biological Rhythms, Vol. 16, No. 4, August 2001, © Sage Publications, https://journals.sagepub.com/doi/pdf/10.1177/074873001129002015
3. The European Organization for Nuclear Research
4. See "Supercomputers aid scientists studying the smallest particles in the universe," by Balint Joo, Oak Ridge National Laboratory, January 25, 2021, https://www.ornl.gov/news/supercomputers-aid-scientists-studying-smallest-particles-universe

Chapter 2
Incoming Storms

Jacob Darbishire, a descendant of the famous Darbishire, whose Schwedischen Südpolar–Expedition in 1901 had found so many celebrated lichens, looked up from the cubital array of data crunchers and simply frowned, shaking his head, as Carlyle, standing next to him, studied the screen.

"That's spooky," he said, and was presently repeating himself, as if iterations might offer unexpected insight.

Jacob sighed. "Yep."

Then, "Savicz, what are you seeing?" A weary meteorologist, Angela Savicz, sat two flights below, in the station's bastion of transiently powered mainframes, monitoring almost continuously from the time the storm had taken such a savaging turn.

"Nothing," came her scratchy reply. "There is a gathering pool of water, I might add. It's filling all the buckets every hour."

"The storm, no other data?"

"No end in sight. I'm looking at projections for the next two months." The late 60s something ashen-haired figure, fully absorbed and obsessive about her numbers, couldn't quite wrap her own head around her extrapolations. "I know. And it's all diverging from the South. Filmy, longitudinal scaffolding, velocities converging."

Carlyle couldn't quite work it out, either. Transfinite numbers, foliated scales, feeder stress, humidity spiraling in unknown magnitudes. "Three days now." *But for the rest of the Antarctic winter?* Normally such heavy snow and ice falls were accompanied by warmer, not colder weather. A limited perspective has its comforts, but none at this time. He remembered that famed line written by Richard Byrd in his book *Alone*, in 1938: "This is the way the world will look to the last man when he dies."

He and Jacob, both men in their early 70s, long accustomed to bare, intermittent bursts of sleep, like the insomnia of wolves, scanned the suite of entangled shadows flashing and accreting quantum computations that could not locate their four comrades who were out there. The others in the station were taking breakfast, in the one

© The Author(s), under exclusive license to Springer Nature Switzerland AG 2022
M. C. Tobias, *The Maiden Voyage of Petrus van Stijn*,
https://doi.org/10.1007/978-3-030-97683-5_2

roomed gym. Those who might prefer to sleep were awake thinking of any alternatives, any at all.

But the cracks in the glazed surface of one familiar terrain had everyone dismally uncertain. Most were resigned to the unthinkable. The back-up batteries for the mainframes were nearly dead. There was no ingenuity left to rectify this electrical dissipation.

Suddenly the chamber was hit by a heavy jolt, both men grabbing hold of whatever they could. A stressed rolling passed through their environs, like the hull of an icebreaker straining against the implacable continent. A deep chasm breaking open in branching thunder. It kept arching in a frenetic space of its own with subliminal cries of an entire ice shelf pressing toward a void. Items fell onto metallic floorboards. Both men continued to brace themselves. Carlyle's heart was pounding. Jacob checked his watch as the shaking was usurped by a series of short thrust vibrations. Such events had been escalating all winter. Then the moveable ensemble ceased, as if cut right through by a tungsten razor.

"Guys?" Savicz called out, her voice in the first of throes.

Everyone at the station was now emerging.

"I tracked IRIS," Basil Heinrich, the base generalist, over 80 years young, exclaimed with steely nerves. He'd lost most of his toes to frostbite during the past two years. "Eight point one. About twenty seconds."

It felt much longer, Carlyle thought. The slow-slip event had been taunting them for months, a continuing seismic tease, sluggishly colliding plates nearer to thee, Savicz had cleverly described.

IRIS was the only viable data stream left within the greater Queen Maud Land network.

"Eight miles out from Breid Bay and very shallow," Heinrich continued.

Carlyle grabbed on the edge of a table as an aftershock struck.

All knuckles present gripped something, waiting. Five, six, seven seconds. Then, "Six point five," Heinrich finally exhaled.

Then another struck.

The ferocity of the icy gales to the exterior did not allow for the differentiation between one shaking and another. The temblors kept coming, first eight seconds apart. Then a minute. The escalating speed of water dripping throughout the many sectors of the station was audible.

Anselmus, approaching 80, a Norwegian stoic who had joined the permanent station team over three decades before, was once considered the foremost geologist and gravimetric specialist on the continent. Now he felt not beaten so much as humbled. He looked to Michiel, the one remaining glaciologist, who was wiping off coffee. His mug had flown out of his hands and shattered on the floor.

"The remaining headland at Breid's Bay will have collapsed," Michiel uttered, a realization glazed over with so many incredulities swimming about in his head.

"The cracks are not cold responses," Anselmus replied. For years that had been the assumption that the sheet was breaking up as a response to surface compression on the backside of ridge lines where dense, cold air pushed against the frozen surfaces, highly pressurized air with nowhere else to go.

Genevieve came up from her room in her pjs. She was a French ornithologist who had become trapped at the station for half her life, a study in human nature ruled by inconsolable dejection. Others felt certain that she had misgivings about ever leaving Dijon. It was supposed to be a single Antarctic summer. She had begun and ended her quest to understand erratic avifauna migrations. The plot and the punch lines of her research had hit their cul-de-sac. Nowhere else to take the data sets, which provided all the clues needed to fully grasp *the situation*, as most referred to it.

"We don't need to be lectured about your fault lines theory," she said, in one of her moods.

"It's no theory," Anselmus quickly rebutted. In his mind's eye he could see the entire half-million square miles of Dronning, as the Norwegians who theoretically once had sovereignty over Queen Maud Land had named it. With its five distinct coastal headlands continuously calving into the churned-up waters of the King Haakon VII Sea. An astonishing set of rapid predicates resulting in an astounding rate of glacial ablation. Half-century before, Dronning was three times the size, but each of the coast lines had lost hundreds of billions of tons of ice. Anselmus had always craved to see Antarctica as it once was, prior to the supercontinent, when all of Eastern Antarctica would have resembled his native Vestland, atop the Hardangervidda plateau, in Norway; with hundreds of forested nine-thousand-foot-deep fjords.

Jacob had turned 72 just days before. He was tired and unshaved. Both he and Carlyle, also a third generation Elisabethan, as the older Belgian scientific base personnel thought of themselves, had lived in the chrysalis of this brewing storm for years, as steadily increasing havoc pitted land, air, ice, the entire microclimates of the Sør Rondane and Utsteinen ridges above them against any notion that normalcy could ever again prevail. It could not. Their immediate surroundings had transmogrified with novel crescendos; magnified in ways that targeted logic, invaded the penumbra of any predictability.

One wanted to openly speculate. But the character of these occupants was such that they simply recorded, noted, studied it for blind posterity. There was no luxury afforded opinions or grave discomfort in cramped, encinctured quarters. There was nowhere else to transmit the news from their blinding darkness. Nowhere.

Chapter 3
The Beryllium-10 Factor

Such had been the circumstances across the continent since Anselmus had first detected the heavy presence of Beryllium-10 in everything—the base structures, rock and sand, ice and snow along with the observable solar flares emitting a roughly gauged quantum of approximately 10^{39}th joules. The magnetosphere had deteriorated into a chaos of *moments* that could be all but divined within a metaphorical Schwarzschild radius [1], as it was known, that shook the world in the size of one's fist. A black hole of tumult and human fatalism. Anselmus lived the physics, could calculate in a flash the necessary constants, initially applied to the mass of stars, but now—by his own brilliant adjustments over the years—to the mass of no celestial object. Rather, to the impact of the Sun on the ice. He mumbled a lot to himself.

There had been a solar hurricane—not a wind. Its strength had normalized a shock wave with heretofore unimagined ram pressure, electron gyro-motion, taking out all communication devices across the planet. At least for many years that much seemed clear.

Save for three weak signals, two of which covered a range confined to Western and Eastern Antarctica. One controlled from pes, which was tied into a network that encompassed the former American, Japanese, Norwegian, and Russian stations. All had been abandoned. The other, at the South African SANAE IV, atop the cliffs of Vesleskarvet, was still permanently inhabited by four or five persons. One of those researchers had evidently been very ill—more cancer than liver. That had been the last reported diagnosis anybody at pes had heard about. No further word on her condition.

The third signal was something else entirely. The one utterly persistent and perplexing glitch in what was otherwise an almost comforting and uniformly irreversible Apocalypse. A repeating beacon emanating from far off coordinates, as if from another planet. But they were unambiguous: 51 12 18 34 N/3 13 34 90E, and there was no way to return the signal.

This also implied, of course, that there was at least one distress satellite up there with a functioning payload, GOES, or METEOR, MSG, or Electro.

© The Author(s), under exclusive license to Springer Nature Switzerland AG 2022
M. C. Tobias, *The Maiden Voyage of Petrus van Stijn*,
https://doi.org/10.1007/978-3-030-97683-5_3

The transmission was a standard 406 MHz but, without any global rescue coordination, it was apparently on a useless timer. As far as the occupants of pes had come to understand, there was no ground station; no world of first responders. If there were inhabitants still sealed away in places like the underground circular collider tunnels north of Geneva at CERN, or the former Homestake Gold Mine inhabited by physicists for decades, at well over a mile deep in South Dakota, it would not matter much. All such facets of *civilization* had long ceased to be relevant.

But Osna had understood at once. The coordinates described the Groeninge Museum, in the heart of Bruges. Her husband had worked there for decades, perhaps still. Petrus was hardly 3 years old at the time Stefanus van Stijn, famed Flemish art historian, took the job. Petrus had been born in Antarctica during Stefanus's 1 year sabbatical to the ice with his wife and son. There he went crazy, not without some dignity, but he simply couldn't stand it. So different was he from his wife. He came to abhor nature's *absence*, like Aristotle contemplating the vacuum, horror vacui, an unapologetic kenophobia, and openly conceded that he could not fathom his wife's obsession with it. He simply missed art history, cobbled streets of the Northern Renaissance.

One could argue that his dreams were mundane, confined to wooden panels or stretched canvas, framed and finite. A quaint but monotonous world far away from the real rigors of radiation monitoring over snow cover, or the timing of growth bands among south polar bryophytes. These were stultifying tasks, as Stefanus perceived them, though he certainly relished the scenery, in the beginning. And he concurred that the company of penguins was never boring. But after a few months, he so longed for his home, and to peer directly into the world of a van Eyck, that the differences between he and his spouse had quickly turned into an obvious marital spoiler alert.

"I don't understand what you hope to accomplish?" he finally admitted, after a mere month on the ice. He was already restless to escape the icy continent. Knowing that he could not sowed grave discomfort, and the beginning of useless arguments.

She had tried to walk him through the early stages of her work with the international genome consortium. But she could not explain the metrics of genomics and why it was so important to get to the common ground of polar genetics.

"Fundamentals of what, amino acids?" he heard himself repeating after her, not having a clue what any of it had to do with the real world of dissipation and sadness. Even when she patiently walked him through its overview, he grew quickly weary of it all.

"Those penguins are going extinct," she told him. "That colony of Adélie you so seem to relish, right over that ridge," trying to drive home her own deeply disturbing grails, given their circumstances, and not least of which, a full mapping of Antarctic chromosomal genome evolutions, as part of a global network of endeavors for all biodiversity. Many of her cohorts had given up in other parts of the world, or so she assumed, having by incremental shock lost all contact. She was lucky to be spared any alternatives on the ice.

The primitive work had started in the late twentieth century with vertebrates like zebra finches, parrots, and hummingbirds. And with the usual model organisms,

Escherichia coli (*E. coli*), fruit flies. But Osna and colleagues had much greater algorithmic plans in store for understanding genome assemblies of literally millions of organisms. It was no vainglory but a true panic, decades long.

Until everything went silent. At which point she would lay awake at night wondering what had transpired. All the plans and preparations. Not on Mars, but across Europe. A frenzied experimentation that had become contagious. First with sheep, goats, then larger mammals. By that time, Stefanus was long gone.

"Of course, as you know, there are magical plants even in your parents' garden," Stefanus had gently reminded her. He was thinking of the sixteenth century Flemish greats, like Joris Hoefnagel and Carolus Clusius. She had promised that by the following Antarctic Summer, she would bring their infant back to Belgium and greatly shorten her polar visits. A promise she fully believed she could keep.

By that time, early into their marriage, Stefanus had put his own theories together.

Reference

1. See Cosmos, https://astronomy.swin.edu.au/cosmos/S/Schwarzschild+Radius

Chapter 4
Absentis Diebus Quattuor

Just two months after Stefanus had returned home, the Sun discharged something altogether new in human experience, and the magnetic poles shifted so erratically as to ensure manifestations that drove people more than half-mad, not just as some saying goes. There was a global panic. Every personal and geopolitical perturbation unleashed. But no one could have imagined that the subsequent chaos would not come to a fairy tale ending; the normal course of some cautionary tales—within a few days, weeks, certainly months, waiting it out on the beach near the *tumid river*. Waiting to be saved by the kind, soft words of an authentic Pope. Rabbi, shaman or other evangelical.

In the words of Mahavira, Maimonides, Gandhi, Tutu, or Darwin's heir, Edward O. Wilson on the Creation.

Alas, the years mounted. Lost causes and statistically dulling static -dim bulbs as the new metaphor for unlit cities, mentation marred by apathy - became a new currency. Pounding weather rendered all rescue operations impossible and irrelevant across the Antarctic. The world of the living hunkered down, wherever they still might happen to be. Or that was the received wisdom. Received, in the sense, that there was actually nothing but ephemeral wave forms of static.

The thing is, nobody knew what, precisely, was happening. Theories abounded over breakfasts at pes, amongst an inner circle of friends, an international who's who of brilliant, eccentric, multidisciplinary Pequod *isolatoes*, as Herman Melville once referred to such living enigmas.

Until there was absolutely nothing to be done about it. No way to get out or receive information. Thousands of volts per mile(s) (that's not much). And then it got worse. Every coil in every transformer was largely incinerated, as all systems connected to any electronic system went 100% dark. NOAA's Space Weather Prediction Center did not see it coming. Pit and pendulum voltage shifts that hit every conductive surface across the planet. The most sophisticated networks were reduced to clumsy DXing between anyone with a Yagi antenna who fancied themselves a ham operator, scanning every conceivable frequency, using center-fed

© The Author(s), under exclusive license to Springer Nature Switzerland AG 2022
M. C. Tobias, *The Maiden Voyage of Petrus van Stijn*,
https://doi.org/10.1007/978-3-030-97683-5_4

half-wave dipole antennas, or complex reflectors. It didn't matter. Calculations of radiation resistance, currents, field expressions, gain and vector, all were impossible to elicit, let alone manipulate. There was no refractive radio wave propagation from the ionosphere; while groundwave, even line-of-sight propagation proved futile. Even the static was disappearing.

These were not concerns of common parlance. People simply had no choice in the matter; no understanding of even the preliminaries. An auto mechanic down the street might have some semblance of the general problem; radio operators struggled to figure it out like one of those grand cosmic puzzles that had dropped into their garages and basements. But, not long into it, there was absolutely nothing to mitigate the new reality.

There were no more skywave transmissions, or E-skip; no distance coverage whatsoever. Marconi could not have imagined any of it. That the earth's same curvature allowing for his three pips of the letter S to cross the Atlantic might work antithetically in a profound, perhaps unprecedented geomagnetic storm.

As the squall worsened, sub-atomic particles were laying waste to eukaryotes far and wide. Any biological cell with a nucleus was at risk. But without any form of communication, science, medicine, every one of the more than 250 million individuals who thought of themselves as *researchers*, were rendered all but solitary, incommunicado.

Presently, Osna, chief biologist at pes, her son, thirty-year old Petrus, whose passion for Antarctic epigenetics he had inherited from his mother, hydro driver Morgane, and lead climber Sanne, had gone missing for a fourth night. No contact with their base, Princess Elisabeth Station.

Remaining station personnel, once, long ago, over 250, now down to fewer than fifteen people in total, had argued various options. But there were none of any real potential.

Everyone had all variously given in to the truth of their plight. No one could escape the pervasive bias of somber inevitabilities that had forsaken pes, and all the other bases with their doomed stragglers left on the last, if quickly fragmenting, continent—decades before. When they're dead, you can't *tell* them that they've died. Somehow this fact alone was the most disorienting mystery of all. Some had died with smiles, and a final *thank you*. But within half-hour, they would get cold, jaws locked open.

Chapter 5
A Clou Within the Frozen Eyewall

At 10,328 ft. (3148 m), Jøkulkyrkja stood out amid the Mühlig-Hofmann range, the highest cirque in all of Queen Maud Land's more than two dozen mountain groups. Osna and her son had climbed most of them, but never Håhellerskarvet, some 800-ft lower than Jøkulkyrkja, and nearly 24 miles away, along the Austreskorve Glacier. The billion-year-old Mesoproterozoic walls of granitoid glowed in the dark. Their metamorphic facies containing colorful lithological veins of garnet and feldspar, gneiss, reddish charnockite towers, bands of rich chemical colors, had first been surveyed nearly 70 years before. But none of the few remaining specialists had ever found the appropriate weather window to permit a climb back up to examine the biometrics. Until now.

The quartz syenite and intense grain of the rocks had admitted countless life-forms, including a fossil leaf which Petrus at once recognized: approximately 45 million years old. But Osna was hoping to find something else up there. In the sharp waxy light in a blizzard-colored pink and gold, the aurora australis having for many years punctuated their world in ways never previously witnessed. The reason was clear. So many moments of magnetic depredation, possible magnetic monopole shifts, disappearing dipoles, voids—as the physicists called them. Most problematically, Jaramillo short term, and Brunhes-Matuyama longer field reversals seemed to have set in (781,000+ years). The jargon had become second hand to them.

A continental instability that was both spinning, decaying, and aggravating all life forms, that was now the language of physicists, geologists, and ornithologists, each. All of Eastern Antarctica was like a gigantic magnetic black hole with zero predictability. Anything could happen at any time. Crazily, that made life less uncertain.

A new global boundary stratotype section and point had been confirmed decades before, not in fossils, but chemical compositions on the surface of fossils. This had occurred along the Western Peninsula, not far from Poland's Arctowski Station. But that was before the unexpected rapidity of so many unaccountable magnetic transitions. A magnet was no longer a magnet. It could just as easily be repulsed.

© The Author(s), under exclusive license to Springer Nature Switzerland AG 2022
M. C. Tobias, *The Maiden Voyage of Petrus van Stijn*,
https://doi.org/10.1007/978-3-030-97683-5_5

These were the peculiar circumstances in which the four climbers now found themselves.

Nestled along a granite crack, Osna found what she'd been looking for. A moss containing invasive multivoltine midge larvae, *Eretmoptera murphyi*. This rioting sprite of minutia had managed to all but out-compete the one endemic, day-dreaming insect on the continent, *Belgica antarctica* (Chironomidae), another flightless arthropod.

But something paralyzing appeared to be occurring, not just among the maverick midges, but various Antarctic avifauna that had for tens-of-millions of years relied upon the earth's magnetic fields to navigate. It was her friend and bird-loving colleague, Genevieve, who had monitored such mixed migrations heading to and from the continent at altogether different altitudes and rotations. Poor Genevieve. It was no secret that she was losing her mind. She identified too closely with birds. Like everyone else, she could scarcely put together a past. In her case, the fact that once she had been married, to a nobleman, she alleged. And used to gab endlessly about two daughters somewhere. Twenty-five years after the fact, she was still going over in her mind the best colleges for them to attend. With far more moribund outbursts on occasion, she contemplated what would happen, to penguins, skuas, to everyone.

Now Osna, too, was seeing a frightening phenomenon occurring in every organism that moved, not just birds. Genevieve's instincts may have been dark, despairing, but they were acutely focused, supra-rational.

Osna had already obtained both first and second instars, and now sought third and fourth to examine the complete metamorphism of the insect. Her counting dish and calibrated ocular micrometer were back in her lab at pes. But for now, she simply scraped and scooped and deposited her prey in half-dozen old fashioned Pasteur micropipettes, fashioned of titanium, not plastic. This was no stratified random sample because it was the *only* sample on this nunatak. Even in these conditions, Osna knew at once what she was seeing. She had never observed such a thing before. Evolution appeared to be in free-fall. Or was it succeeding at new levels? Failure, success, change… It was all the same in the eyes of someone, a large vertebrate, who would never again take part in the saga.

Her years of study on the Continent had revealed that natural selection was accelerating in real-time in a myriad of organisms due to the quantum light harvesting pigment proteins, otherwise known as the Fenna-Matthews-Olson Complex [1], all under new types of harsh, environmental drivers. Petrus helped his mother get the dropper with its fresh contents into the carrying set within her pack. She would test the chlorosomes [2] under X-ray spectroscopy when they made it back to base. Osna was well along the way to proving that a changed world had exerted enough stress on insects, plants, and birds, among others, to have transformed the prospects of adaptation and resiliency. If there were to ever be a *normal*, these protein complexes would be the ones that championed a new epigenetic quotient. Although in her sleepless late nights she could not envision what the victor emerging from the biological chaos might be. What would it look like? What would it do? Was it all a beleaguering dream, the addled conceptual substrate in-between the morphine drip of their over-medicated scientific bubble?

Petrus had been mentored by his mother in all things totipotent (unfettered cell differentiation), one of many of Osna's hopes for a future. It was a tomorrow teeming with covalent modifications. The activation of certain genes specific to conditions of stress. The scenario was first laid out by the early twentieth century philosopher, James Mark Baldwin, a distinct voice in the biological arena of the so-called modern synthesis [3]. Osna had been partly persuaded by Baldwin's organic/functional selection theories during her doctoral dissertation years, when the so-called evolution wars in scientific circles were grasping to account for mass extinctions resulting from the combination of so many human inflictions. It was Baldwin who had provided both the ontological framework, and common-sense narrative to perhaps best gauge the future of biomes based upon an organic developmental phenotype independent of logarithmic or exponential bases.

Osna had believed that Baldwin's "Effect"—the hopeful spin on ecological nurture [4]—across real-time generations was the only explanation for the readily observed empirical fall-out, as humanity's stress inducers multiplied upon all other life forms. It was only logical that species would react in ways that brought back their ancestral, genetic pertinacity. Or was it?

But these were only the most obvious forms of biological calculus. Osna had applied theories of mounting stress to multicellular intergenerational floral assemblages throughout the mountains of Antarctica, noting that both hair grass (*Deschampsia antarctica*) and Antarctic pearlwort (*Colobanthus quitensis*) were, in fact, not the only two endemic plant species on the continent. But that stress had forced hundreds of others to become so, breaking up traditional genetic boundaries on the basis of genotypic behavior that had become congenital in species after species. It was big news, she thought, but possibly she alone. It added entirely new dimensions to the nature/nurture debates once thought to have been resolved by sociobiology.

As for Petrus, he had become skeptical at a precocious, early age. Too much had gone wrong, from what his limited experience in the furthest corners of catastrophe had conveyed, to be convinced of any one scientific promise, let alone those theories that were washing away. Out to a nameless, borderless future, in a nameless time. No theory could help you in the cold, not this cold. One just wanted, at least some of the time, to stay alive. He inhabited a lonely, incommensurable era of such shattering silence, few contacts, a purposelessness. But his vigor and stubbornness were a power all their own, defying the cliches.

Osna insisted to Petrus that one—they—had to sustain some order of *faith* if you were to persist in the study of seal wallows, glycerol concentrations in Antarctic micro-arthropods, or the sampling of hydrated microbacterial colonies.

"It doesn't quite rhyme, does it, mother?" feeling swept by nonsense.

Osna knew her son longed for the home he'd never had. Not this. Nor did she have any answers for him. If she did, they had fogged in her mind years before. Recriminations wallowing in what were to be the final days of narcissism.

Petrus did not, exactly, buttress any singular suasion. He was perturbed by his protracted isolation, but since he only knew the non-Antarctic world through holographic discs and books, his finitude was equivalent to some kind of perverse orphanage in space. Disc time per occupant of Pes was down to 15 min per week, on an as-needed basis for research, on account of the dwindling battery packs.

But books were there in the annex dome in profusion. Petrus had examined much of the work done by such notables as Ando in 1979 on bryophytes, Lindsay and Brook on lichens, Pickard on the "Vegetation of the Vestfold Hills" and Engelskjon on various botanical aspects of Dronning Maud Land in general. He was particularly intrigued by the creatures inhabiting Gjelsvikfjella and Mühlig-Hofmannfjella, which Engelskjon had compared with existing data bases as of 1985 of other Antarctic flora. The changes in well over half-a-century (pre-adj) of enormous environmental calamities had posed questions about life forms never even contemplated prior to this time. Phytoplankton biomass counts, measurements of the Japanese BIOMASS program, "decomposition of sulphur-containing proteins by putrefactive bacteria" [5].

Once, well before the birth of Petrus, Antarctic biology had seemed so straightforward.

No one working in the field had time for such questions as, where are we all headed? There was the simple answer, of course. But then there were the *real* answers, the ones no one could anticipate. But a century later no one dared to speak of what was really happening, not after so many years in its ungraspable maw. No talk of a mass communion or ritual suicide. Nothing quite like that, though the intimations were clear. Even their little hamlet of survivors lived only barely by rules and cohesiveness.

Some openly meditated on God. Others took a different side, the accumulating filth in the communal bathrooms. A broken door that could prove fatal. Each had to choose a corner as the madness, like any plague, spread. Even from the bottom of the world. Faith was a supreme liability should you choose wrongly, Osna would argue. Petrus never understood what she meant. She never explained. In any case, words had begun to reveal their very absurdity. Sentences, discourse, even the most benign discussion yielded to an obvious exhaustion in the air.

Petrus also knew that Osna had aged poorly. There were times when he worried not only about her *scientific memory*, but her thinking patterns in general. There was nothing that could be said or done. It was quite easy to speak of big ideas. But no one wanted to hear about the same old food at mealtimes. Most activities had once been convivial. Not anymore.

Presently, such academic concerns as natural selection and phenotypic change held little interest to the four-member team that huddled for its lives, trapped hundreds of feet up the wall, pinioned on a ledge in near zero visibility. The shape memory wires (aka ropes) had become stiff. The winds had the four expeditioners backed up against the red cliff. Their re-breather masks had uniformly broken down with spalling hail shattering every inch of virtually unbreathable air. A fire in the lungs.

Petrus was the strongest member of the team, and by far the youngest. Now, he was afraid.

References

1. Rémigy, Hervé -W.; Hauska, Günter; Müller, Shirley A.; Tsiotis, Georgios (2002). "The reaction centre from green sulphur bacteria: progress towards structural elucidation". *Photosynthesis Research*. **71** (1–2): 91–8. doi:https://doi.org/10.1023/A:1014963816574. PMID 16228504.
2. Oostergetel GT, van Amerongen H, Boekema EJ (June 2010). "The chlorosome: a prototype for efficient light harvesting in photosynthesis". *Photosynthesis Research*. **104** (2–3): 245–55. doi:https://doi.org/10.1007/s11120-010-9533-0. PMC 2882566. PMID 20130996.
3. See Mayr, Ernst. *Systematics and the Origin of Species from the Viewpoint of a Zoologist*. New York: Columbia University Press, 1942.
4. See *Evolution and Learning: The Baldwin Effect Reconsidered* edited by Bruce H. Weber and David J. Depew: Cambridge, Massachusetts, The MIT Press, 2003. See also, e-Annals of Stress Photobiology, "The Last Scenario for Photosynthetic Efficiency in Active Aira Caldera," Vol. 18, pre-adj, 2029, Moscow, Russia. Doi: Inactivated.
5. See "Decomposition Processes in Maritime Antarctic Lakes," by J. C. Ellis-Evans, in *Antarctic Nutrient Cycles and Food Webs,* Edited by W. R. Siegfried, P. R. Condy, and R. M. Laws, Springer Verlag Berlin, Fourth SCAR Symposium on Antarctic Biology, 1983, p. 258.

Chapter 6
Doomed Logistics

Sanne shook his malfunctioning LIDAR, then noticed a particular band in the rock. He'd been on the reconnaissance of the range two years prior. "I remember this. Okay. That puts us about 200 ft above the glacier."

Petrus helped his mother who fumbled in her effort to attach the self-arresting descender. Their thermal reactant gloves, climbing suits, ice-repellent boots, kept the cold only partially at bay. Moreover, breathing, even in short half inhalations, was painful and spasmodic. If they could reach the base of the cliff, where their two turbojets had been secured on the glacier side of the small bergschrund, they might stand a chance.

They were 360 mi from pes. Four days before, they had aquaplaned over the glaciers at a safe 50 mi/h. They had camped the first night. By noon the following day, as the storm roared in, Osna had collected her crucial samples. Now, the four of them had to get into the bergschrund and out of the storm. They had never been exposed to such winds or temperatures.

Sanne drilled two stainless steel power bolts, set up a belay of multiple twin gate locking carabiners, and, saying few words, urged Morgane, then Petrus, and finally Osna, off the wall, down the few hundred feet of wire which he had threaded. He would go last.

The abseils went flawlessly as they were descending the north face, which gave some respite from the blizzard coming from the backside. By the time all four stood on the cornice, however, they were face into other angles of the super-tempest. Sanne planted his ice axe, tossed the wire into the crevasse, and mounted a rapid recce. It was deep. Too deep. There was not one ledge in the shocking blue darkness.

Osna looked anxiously to her son.

"Mom?"

"My toes are frozen solid."

Petrus shined his headlight directly on her boots. Ice was layered thick. The micro-cellular heat exchangers were malfunctioning. Nothing to be done about it.

He saw a fear in her face he had never witnessed. That was his *real* mother.

© The Author(s), under exclusive license to Springer Nature Switzerland AG 2022
M. C. Tobias, *The Maiden Voyage of Petrus van Stijn*,
https://doi.org/10.1007/978-3-030-97683-5_6

Chapter 7
The Evacuation Attempt

Morgane had stretched the aluminum ladder across the seven-foot-wide bergschrund and was checking on the vehicles, as Sanne ascended back up from the chasm. "It's not going to work," he said, teeth chattering. They had thought that, perhaps there might be a ledge on which to escape the full brunt of the storm.

Petrus noticed ice on Sanne's boots as well. He shouted over to Morgane: "We have to go!"

They'd be heading at a cross current to the storm, two per vehicle. The yellow halogen fog lights helped only at low speeds, which they could ill-afford. Three feet of snow had fallen, or more, so they all agreed that the hydroplaning was safe at 80 mi per hour, with the combining of lidar, sonar, and radar to gain some mapping of obvious hillocks or unknown crevasses before them. The problem was the size of the incoming hail at such speeds and angles. Their masks were fogged, one of them, Morgane's, cracked.

Petrus helped his mother into their vehicle, while Morgane and Sanne pre-heated their own 600-pound turbo machine. Winds had not abated, the ice storm was blinding. In all their collective years in Dronning, such weather, even in the heart of Winter, was rare.

There was no way to sort out provisions. Winds made it impossible to extend an arm. To stand firm was difficult. It was enough just to keep one's head low, bodies covered to some extent in thermal blankets. Navigating was up to Morgane. He had instincts and still functioning technology.

Both vehicles roared into the blizzard, heading on a northeasterly course. Eighty miles per hour, much of that speed above the ice, two, three feet into the air, bounding back, then bouncing back up.

"It's going to be alright," Petrus shouted, without turning.

Osna, holding her head as low as possible, and gripping both sides of their open vehicle, said nothing. She knew what was in store. She'd been caught out in lesser storms before. But they had not factored in 80 mi/h wind chill on top of everything else.

© The Author(s), under exclusive license to Springer Nature Switzerland AG 2022
M. C. Tobias, *The Maiden Voyage of Petrus van Stijn*,
https://doi.org/10.1007/978-3-030-97683-5_7

She thought back to moments in her life. Clutching to an ER bed handle at pes where she'd given birth to Petrus, the first of their generation to be born on the continent. She blamed herself for her son's extraordinary situation; his total innocence. If she had only listened to Stefanus. There were numerous chances to have gotten their son back to Europe, off the ice, had she only insisted. Her stubbornness was abetted by the young Petrus' budding proclivities for the biological sciences. And then, 1 day, just like that it was too late.

Her husband's family had come most recently from the Netherlands, where there were several cousins named Petrus. All but one, Piet, of conservative stature. Piet was a double-down drug addict who died in his 20s. In some ways, Stefanus would say later on, when he and Osna could still communicate with one another, he was possibly the lucky one. Going out in a comfortable haze of heroin bliss before the world turned.

Snow conditions had made for a bizarre and treacherous labyrinth. The old fern snow comprised a dangerous degree of solidity exceeding 1200 hundred pounds per cubic meter. Hitting it head-on, amid the mazes of enormous ice megadunes and barchans, penitents and sastrugi, left a blinding after image, mirages that were real. The storm bursts had metastasized into the graupel Petrus had never encountered, supercooled ice droplets and masses of polycrystal that could leave more than mere puncture marks on one's face, especially at such speeds.

Then Petrus realized that there was no fog light before him. The other vehicle had vanished in the superstorm, whose northerly gales were accelerating the speed of Petrus' hydroskimobile. Either he was off course, or something else had gone terribly wrong with his two colleagues. Now the speed of his machine was reaching a point of no control.

"Scott Base, Antarctica," © By Craig Potton, Courtesy of Craig Potton

Chapter 8
Lone

Osna was trying to say something but the roar of the fuel cell stack-powered open conveyor, against the hurricane force powers all around them, muted her voice. Petrus was working desperately to hold to a traverse that avoided incoming obstacles, none of which were recognizable.

Then it happened. Like an explosion with ripple effects seizing thousands of square miles of geometric space, sucking in all light, as if the continent had become a black hole, then expelling the energy waves.

If an observer with unobscured light were to have viewed the collision it would have looked almost like an Olympic event charged with a thrill. The hydroplaning had morphed into true flight. Where there had been a relatively level ice cover now was unstoppable cliff. The physics of their trajectory mattered little, except to say that at 80, possibly 90 mi/h, the 600-pound vehicle with its two adult occupants was now sailing in an uncorrectable twist off Queen Maud Land. The continent had fractured, the coastline breaking inland off the ocean, consuming all of pes—its nearly 5000 cubic meters of interior living spaces—which once sat indomitably more than 120 mi, then 75 mi, and then 8 mi from the King Haakon VII Sea. The collision on to the pack ice was muffled by the bewildering clamor of confusion. Osna was killed instantly. Petrus flew, hitting headlong into a concrete-thick megadune.

When he awoke, his helmet clearly having saved him from what he could feel to be a severely bruised forehead and neck, he saw that less than a mile or two away was pes, a portion of it on fire. The blizzard was overhead now. He knew it was swirling upwards from the fixed ice walls that held back ocean from headland. It must have been Breid's Bay, where such walls were many hundreds of feet high. The storm was moving out over water in the direction of South Africa, over 2600 miles away due North, across the Southern Ocean. The mosaic of disasters consolidated into an unspeakable chaos, in his bruised head.

M. C. Tobias, *The Maiden Voyage of Petrus van Stijn*,
https://doi.org/10.1007/978-3-030-97683-5_8

27

He fumbled with frozen hands to try and extract his mother from the vehicle but her legs were trapped beneath the PEM stack of fuel cells and other ruptured metal equipment. She was lifeless. If, indeed, there had been a pulse he could now feel nothing. He searched beneath her helmet and parka. Upon removing his gloved hand, it was saturated in blood. He pulled up Osna's turtleneck and there saw the gash. Her neck was grotesquely twisted and it was clear, blood gathering in her mouth, that internal injuries had been fatal.

He tried again to lift her from the wreckage, but it was not to be. Such impossible cold speaks. He had to get out or die. How does one die when as yet he is alive?

He was now alone.

Chapter 9
Salvaging

Petrus had been trained from the near moment of his birth, and before, to be what some might describe as a *practitioner* of Antarctica. He learned early on how to function. Few events surprised him or caught him unprepared. His parentage was strongheaded, though he really did not remember his father, Stefanus, very well. Two years or so of virtual communication, before such options were lost. They had not seen one another in over 29 pre-adj years. Petrus had long ago lost track of everyone and everything outside of Antarctica, where he quickly grew out of child-hood and, by the age of nine, was a working biologist, with no other prevailing interest.

Because the years rolled by, it never occurred to anyone, other than Osna, that Petrus van Stijn was the only human being to have been born and raised on the Antarctic Continent, never having left.

Now, for the first time ever, the fact that the world had changed smacked him unmercifully. Petrus reached the ruins of pes, crossing ice that was thick but inde-scribably marine-like. He could hear the thunder of crashing waves, the sound of all the birds he knew well screeching and calling amid the clutter of some continent-wide transition that had catapulted millions of pounds of concrete and steel, the entire Princess Elisabeth Station and her occupants, tens-of-miles out over the edge of the continent, where the bastion plummeted, or was aggregated into pack ice by some unimaginable concatenation of forces.

Multiple strobe lights and eerie-sounding alarms had been triggered and would not be stayed. The lights circled in 60 s circumferences highlighting a total stillness in the air. The storm had passed over and now stars shone in a hairy constellation of brilliant lights, falling upon the stunned ambience of crying birds and of the unnam-able crunching of chunks, sheets, thirty-story walls of ice churned up in eddy-spiraling seas.

Petrus rummaged and dove into ruptured tunnels, open rooms, and still unbroken chambers, looking for any survivors. He also sensed the strangest sensation of all: The entire edifice of ruin was moving. It induced a quasi-nausea in his gut.

© The Author(s), under exclusive license to Springer Nature Switzerland AG 2022   29
M. C. Tobias, *The Maiden Voyage of Petrus van Stijn*,
https://doi.org/10.1007/978-3-030-97683-5_9

As he made his dreamlike, nearly becalmed survey of his home, pes, one body after another emerged from under various rubbles. The shattered remains of laboratories—sequencing rooms and specimen depositories, of private studios and their mixed corpses.

The death toll was 100%. A ubiquitous pall of everlasting injury attached to his eyes and his heart, indescribably. As if in being the lone human survivor of some Armageddon, he and the birds would be in a unique position to render commentary. No matter what it might or might not be worth. That was all he had left to cling to. For all that, he experienced no fear, welcoming whatever end was in store.

I can do this, his hollow insides echoed.

Chapter 10
The Plight of the Emperors

The station library was up in flames and as he gazed with wonder upon the scene of desolation he remembered back to years and years of reading the exhaustive scientific literature. Hundreds of bound studies. But also Ovid, J-P. Sartre. And his mother's favorites, the refrains—particularly those aimed at saving souls from very real disaster—of Anna Bijns, the sweet and simple poetry of Guido Gezelle. He'd skimmed the art essays by Max Rooses, and, of course, Stijn Streuvels, Gezelle's nephew and perhaps the most prolific and restless of the twentieth century nature writers in Flanders. Streuvels was also a "Petrus" (Franciscus Petrus Maria Lateur) who lived until almost 1970, losing the Nobel Prize for Literature to Pearl Buck. Most of all, Petrus admired the Count of Belgium, Maurice Maeterlinck's *L'Intelligence des Fleurs*, a masterpiece.

All melting amid fire and ice, embers drifting by Petrus in a sensation of great clarity, of destruction that all heralded some higher calling. It had to if any of it were to have the least meaning. But there was no way to calculate such effects.

He managed to pull two bodies from the flames, but then an explosion deep beneath the station sent him flying. His ears were dumb against the frozen darkness, though it was well into the afternoon, he knew, by the dulled, smokey register of sunlight on the horizon's slate grey edge. Dumbness of senses but not of resolve to find out what had taken place. Compromised eardrums. Smoke inhalation. Cervical vertebrae knocked out of whack. His left arm was weak, nerve damage in his neck more than likely.

He sought out the communication chamber, only to discover that it, too, was on fire, utterly toxic fumes. He covered his face. There was spillage all around. He ran, to the extent possible, fearing another explosion. He had less than 10 s to prove his instincts.

Cannisters of gases going off. Chemicals spewing into the surroundings. Although pes had been the first emissions free station on the continent, there were countless flammable objects and gases that could easily and had, ruptured. The station's food court had been crushed into various corners, but there was food

M. C. Tobias, *The Maiden Voyage of Petrus van Stijn*,
https://doi.org/10.1007/978-3-030-97683-5_10

everywhere. He started collecting muffins, cakes, sandwiches still wrapped, tomatoes, and other products of pes' exceptional greenhouse only partially now in-tact, though what looked like a still functioning kitchen. The strips of Brite Lab LEDs had not been damaged, working off a hydrogen fuel cell array that had escaped destruction.

A skua flew through the kitchen, picking up an apple from the floor and shifting its low-lying flight pattern instantaneously upon nearly hitting a dead Alsatian shepherd, Max, one of his legs still jerking, an imponderably heavy steel fixture of lighting having crushed him instantly. Petrus had loved Max. He stroked the fine hairs on his stomach, but the head was obliterated.

Far off Petrus heard emperor penguins, a cacophony. He knew exactly where he was: the colony of emperors—more than 9200 monogamous pairs—occupied a very distinct stretch along the Princess Ragnhild Coast, constantly in flux. Even as a child the ice shelf was disintegrating, water from the inland glaciers pouring into the ocean, perpetual waterfalls. He used to wander among the penguins (several cohabiting Spheniscidae species) for hours, whenever possible. He knew many of them individually. They knew him.

Once, he had been marooned by weather with the emperors. Their standing phalanx of measureless feather battalions helped keep him warm in a blizzard.

Then their numbers started demonstrably declining.

Petrus dug his way through debris into the upper passageway that led to his own bedroom. Amazingly, his bed remained unscathed, pillows and all. Though his entire cabinet of curiosities collected since childhood containing little meteorites, fossils, colorful feathers, books, a photograph or two from Belgium of grandparents who were color faded ghosts had collapsed violently against a wall. He spent the following hours cleaning up the room, seeking out essentials in his bathroom. Until, in a sublime sort of numbness, he figured it was all back to some kind of normal. He would continue to live there and make do.

Now Petrus moved his way back outside, having surveyed that which could be salvaged, and that which was permanently to enter the great fog. He would determine from the seismograph the extent of the earthquake. The exact data was not his area of any expertise but showed up with some obvious soundings. The fissures in the ice pes' scientists had been studying for half century had finally succumbed to all the building pressures of a *moment*. In the dark hours he could not know that the whole northeastern side of Antarctica had come apart. What he did sense was the movement. He was indeed transmigrating.

One ice sheet tens-of-thousands of square miles in expanse (he had helped scout much of the cracking) was, in fact, moving out to sea, further and further away from the mainland, in a northerly heading. It didn't take him long to recognize the signs of what had happened.

Putting it all together, he reasoned, so large an *iceberg* was not likely to melt anytime soon. It was, in essence, a floating island and he was now its sole human occupant.

I've got to bury her, his mind hastened. *To retrieve her samples from the wreckage.*

But within half-hour as he wrestled his way through the maelstrom of ice, he saw the problem. Already his *island* had floated some distance, at least a quarter of a mile, from the fresh ice walls newly exposed to the sea, the new mainland, as ferocious swells, tsunami heights, thundered against the towering, jagged spires rising towards the stars, angry spume spalling 200 ft into the air, higher still. The beauty of it all.

But each night was sleepless. He could not—and was determined never to—forget the unshakable inexplicability of eternity. It was the eternal fact he would never see his mother again: beloved, impossibly beautiful and rare, able to be generous, cranky, headstrong, incisive, forever in the thick of new discoveries, bemusing, guileless, and powerful beyond all versions of its meaning, at the same time. Her last 48 hours had been revelatory, then hellish beyond all doubt. Her death had been instant, of that he was certain. But this was the infinite so many speak of. The infinity that would ensure he never saw his mother again. It so haunted him as to leave him doubly a castaway, triply forlorn, on his ominous sheet of splintering ice.

It appeared that Petrus, they, all of it, was traveling slowly but without respite. Some enormous tabular icebergs were known to move 4 mi\h out at sea, depending on a number of factors. The island, as he would come to think of it (more like an icy archipelago), was traveling by at least half that speed. To the South, a thunderous sea, impenetrable. He knew where old hypersonic zodiacs were stored. One remained unburnt. But there would be no way for him to ever reconnect with his mother. She was not on the island. The forces would quickly bury her on that fading mainland of opaque, interminable consequences.

Antarctica was slowly receding behind him, as he and a number of beloved emperors, Rock Hoppers, King, and Adélie penguins drifted away. As death, he thought, entered him for good. It was not something outside, it was inside now.

Chapter 11
Journeying

Instinctively, his first task was to try and do something with the corpses, over fourteen bodies, including the few pets, like Max, a hamster named Leo, a seal pup only recently rescued, and two friends of Max, Sakhalin Huskies of legendary stamina, their remains legible by bits of unburnt fur.

Herewith, the entire congeries of his human world. Petrus knew mourning; bird by bird, seal and squid, a blue whale consumed by dozens of killer whales, and the like. All of whom he had known and seen vanish in the murderous veils of Antarctic brutality.

Nature is no romance. He had few illusions. And from what he knew of his father's pursuits, most of the Renaissance art was concerned with normal people praying beneath a Christ on the Cross. That was human nature responding to God's nature. No elusive code, no algorithm could enable scientists or anyone else to solve that obvious condition. Medieval sacraments could never have anticipated the total collapse of the Atlantic Meridional Overturning Circulation, which predated Petrus' birth.

He had watched scientists perish from pneumonia, septicemia, a burst appendix. But now, it was a different blow that invaded his otherwise calm apprehensions of the future. No one ever *deserved* to die. There should not have been a god who— under any circumstances—approved of vengeance, he thought. Osna had raised him to believe in one god, the earth god. Though her typically understated meditations never ordained one thing or another. She was no affront, neither demanding nor refraining. Osna's own parents, she always described, had been dedicated to science, without contest, debate, or discrimination. She had gotten on well with both her parents, who had, in turn, described the horrors of World War III. Of multiple viral plagues fabricated in labs from whence they were deliberately liberated. And of World War II that *their* parents, uncles and aunts and surviving cousins, had told to them. Of times dominated by total mob chaos and betrayal of neighbor by neighbor. Of the marches and unspeakable genocides in ovens then trenches. And of aerial bombings. Scenes that now passed along to the great grandson. The family,

© The Author(s), under exclusive license to Springer Nature Switzerland AG 2022
M. C. Tobias, *The Maiden Voyage of Petrus van Stijn*,
https://doi.org/10.1007/978-3-030-97683-5_11

like so many in their time, had been dispersed, some to England, another to Canada. Three to Montenegro. All were committed to the right things, and mostly, she said, idealists to the core, despite cruel reversals, which each family member, like all humans, must necessarily hold tight to their hearts. The condition of all apocalyptic diaspora.

She often described Stefanus to her son. How he was no monarch, but the softest of males, she mused, remembering back to what she more generously remembered as his dominant reluctance to be condemned to the cold. It was a fear as powerful as vertigo, Stefanus had explained. Northwestern Europe in winter was bad enough. He feared the loss of the very rudiments of a social milieu which simply was not possible near the south pole. Months of sub-zero weather were the real zinger for him.

Osna had been introduced to him by a mutual friend, a professor at Ghent University where Stefanus was finishing his dissertation on the back-story of the "Madonna with Chancellor Rolin" [1] and the dialectics of perceived beauty in early Flemish works of art. Hugo van der Goes' altarpiece at Berlin, diptychs by Memling, panels from the workshop of Gerard David, these were his fodder. He was a dreamer and an historian. Trades his parents happily underwrote. It was in the van Stijn's DNA. To paraphrase Kierkegaard, all subjectivity could become truth with enough passion.

"He always said the van Eycks, or Petrus Christus would have paid me to do my portrait," Osna reminisced. It was the greatest compliment.

Names that conjured very little in Petrus' mind, especially as he grew up knowing only the harsh truths of his surroundings. Now, that included the disastrous law of violability of all things. His failsafe refuge was no more. Even the idea of it as having once consumed lifetimes, a mere ember.

Reference

1. Perhaps painted by both van Eyck brothers, Hubert and Jan (John). See, The Van Eycks And Their Followers, by Sir Martin Conway, MP, E. P. Dutton And Company, New York, 1921, p. 64.

Chapter 12
As the Last Continent Vanishes

As the distant remains of the continent grew remote, haze bedimming darkness, lit up only in the resplendent, frightful late nights, Petrus took advantage of the crystal clarity of greatly improved weather to drag, then bury one body at a time. After each brief ritual, he would return to the ruins of pes and work to re-assemble some semblance of order, re-engineering power as possible, securing one of the functioning freezers, and taking stock of the meager food provisions. The power was nearly gone, the last of the fuel stacks depleted, the batteries on their final weakened charges. Petrus now inhabited a near vacuum in terms of electrochemical potential.

For many years, occupants of pes had been on their own, having long before devoured remaining supplies from the outside. Initially, there had been the occasional trades arranged between base personnel across the Dronning network. Such gatherings grew less and less interesting as all the researchers found their respective caches dwindling in like fashion, so that the Japanese were unlikely to part with any more dried seaweed, or the Koreans their kimchi, Russians their pelmeni or Vodka. As for the Belgians, they had lost the bulk of their homegrown chocolates years before. However, they could still produce the thick-cut floury Bintje potatoes, deep fried, vegetable, no beef tallow, but served with the official Mayonnaise, as by order of Baudouin (the King) in 1955. That always impressed scientists from other bases.

By the time Petrus had reached his early teens, the baked breads, or sledging biscuits, mixed with whatever was fresh in the greenhouse was his standard regimen. Some adopted the dire circumstances to fashion hoosh recipes in which the pemmican encompassed seal meat, or squid or krill harvested from their summer and autumn spawning hotspots. But the *krill base* overall had been significantly depleted by UV radiation. And then, in due course, the weight of opinion shunned any harm to any animals. A ban was put in place by vote against taking any wild, living thing, other than krill. Then the depletion of baking powder and raisins for making yeast left a most displeasured company of men and women. Rum was sought with desperation from other bases. Cash payments in unheard of sums, or

M. C. Tobias, *The Maiden Voyage of Petrus van Stijn*, https://doi.org/10.1007/978-3-030-97683-5_12

debts, became common. Until it was altogether clear that debts could never be repaid. The little democracies of Antarctica fell prey to the human natures at nearly 60 bases.

Still, the remaining occupants of pes were far better off than most. Their greenhouse more than sufficed for fruits and vegies, from spinach and tomatoes to nectarines, cherries, grapes, and apricots, regardless of season. There was no scarcity of water, for as long as energy to the stove from a fuel cell stack remained. As Petrus rummaged through the remains, he found that a few of the metal hydride storage cylinders for the hydrogen powder, and the solid oxide electrolyzers, had not been destroyed in the fire. And there was still an old antique stove with a store of natural gas, and a drawer full of matches for back-up. One way or other, he'd be able to melt water. And he knew how to collect rain easily enough.

Now Petrus had to rely on altogether singular skills: gardener, the sailor, survivalist. All the natural sciences in the world had been eliminated from his list of priorities. Decades of research, obsolete. The collective knowledge base of every station in Antarctica, useless. Only the actual experiences, and whatever precise memories or intuitions they had fostered—abetted by pieces of near junk requiring expert tinkering—were anymore worth anything.

He didn't quite realize it yet, but for many years he'd been in training for this, like the original astronauts, or polar explorers. Roald Amundsen, Norwegian member of the 2 year Belgian Antarctic expedition in 1897, went on to reach the South Pole in December 1911. He later perished in a plane crash in the Arctic, his body never recovered.

"Cape Royds, Antarctica," © By Craig Potton, Courtesy of Craig Potton

Chapter 13
A Floating Archipelago

The ever-present rumble and groaning of his island made sleep or any decisive thoughts difficult. How large an ice sheet was it? It would not presumably follow the normal rules of an iceberg, because it was no berg, but an entire shelf, a moving mass without ascertainable depth, not that it mattered except that he could not grasp the speed at which he was traveling.

From his imprecise triangulations out away from pes, where he dragged and buried corpses, he could see that waves were lashing the strewn fields of ice within two miles of the remaining habitat. Enormous swells breaching the fracture lines, spume erupting now like systematic geysers along the entire visible length of the jagged edges, hundreds of feet high. He kept watch on the comings and goings of the emperors (who held a special place in his heart), the skuas, petrels, the numerous gulls, terns, albatross, and jaegers. He knew them all and understood that while they were implacable realists, these birds were also victim to the ongoing moments of global magnetic chaos. He could not depend on avifauna to know where they were heading, or which direction *he* was moving. (Rainbow trout were classic examples of gender predicating entirely different migration patterns; or the "fasting endurance hypothesis" as applied to northern flickers, a woodpecker in North America) [1]. Had pes been situated far to the West, along the Antarctic Peninsula, he would certainly have known when, and at what approximate speed, the island crashed into South Georgia. But there were no sub-Antarctic land masses to his knowledge immediately due north of Queen Maud Land.

Through the violent open ocean he traversed. As if in tandem with leopard, Weddell and southern elephant seals, the blue whales and others who for millions of years had guided themselves beneath Sagittarius, and the Large Magellanic Cloud, through miles-wide behavioral masses of zooplankton, under and above the multi-dimensional worlds of ice and mixing currents, North, South, circumpolar superpositionings and molecular whorls.

© The Author(s), under exclusive license to Springer Nature Switzerland AG 2022   41
M. C. Tobias, *The Maiden Voyage of Petrus van Stijn*,
https://doi.org/10.1007/978-3-030-97683-5_13

Between earth and the lightyears above. Through self-reflecting gazes, mirroring shoals, strange beasts of every color. These were the multitudes of all the transmigrating spheres that lent preternatural life, endowed an endless splendor in brains, feelings and the sheer orientation that had made it, to date, all workable forever and ever. From his moving perch Petrus watched as the currents segued from one dynamic to another, dragging from their customary routes and realms the last vast biological commons throughout the Southern Ocean.

He spent his time observing. Counting and re-counting things, for no good reason. Trying to recall words of various languages he'd picked up from all the international travelers to pes over the years. Mostly he slept, working through kinks in his neck. He sensed that his cervical 1 and 2 vertebrae atop his spine had taken the bulk of blunt trauma, probably fractured. But they were hairline cracks or he would not have been able to drag bodies, he reckoned. Either that, or he was more immune to pain than had ever been explained to him.

Petrus was losing a pound or more every few days. Stir crazy and frightened, his only pleasure was vested in the practical appetite for staying alive by whatever means. *Just keep breathing*, he repeated.

Emperor chicks had already been born, some prematurely, and the parents were greatly agitated. The young were grouping together for warmth -adults containing as much as 188,000 feathers per bird. But many of the adults were already heading into the waters for food. Their cycles had been disrupted, though during the past years this had been happening more and more. Now, the situation of ice mass loss was incomparably more dramatic than ever before. Down was withering on the young, not gaining. Even subdermal fat was declining in penguins. Their epic was just beginning, the Anthropocene even exceeding the unimaginable ordeals described in the literature by such past luminaries as Apsley Cherry-Garrard in his 1922 memoir, *The Worst Journey in the World* [2].

Given that their 70-million-year evolution, arising during the Paleocene, and a steadfast sixty-day period of incubation, succeeded by all the normal chores and itineraries of parentage, the emperors had inculcated a level of cumulative wisdom. The males were in charge, at least in emperor penguin life cycles. They sat/stood upon an egg with a renowned durability and tolerance. Unlike, say, the very picky albatross, emperors needed no nest. It was different with the Adélie penguins, one of whom, an adolescent, had strayed from its creche and cried continually. It sought out any form of companionship it could with Petrus.

In the days after the completion of the burials, Petrus wandered in finely measured sorties to try and grasp the dimensions of his floating island, and count the emperors, looking for any individuals he might recognize. Eggs had been demolished. Males were in no small disarray of frantic grief. There were fewer than a hundred birds. He thought he knew most of them. They certainly remembered Petrus. The scene was one of zoosemiotic inconsolability, feelings flushed with that doom no commiseration can touch, although it tries. The birds displayed a mournful stoicism in their eyes whose haunting, impenetrable gaze appeared to transcend their deeply vulnerable futures.

He could not, of course, count the ways, imagine all the memories and primeval scenarios that evolved in these birds, whose tenderness vied with a desperation that was simply heartbreaking. It was there, on full display in their wise old eyes. Again, he thought: *Biology is horrible*. Nature raw is nature awry. What did he expect?

One night, 11 days after the disaster, Petrus lay in his bed thinking back, not forward; and finally succumbed to the impact of Osna's horrible end. He knew the brown skuas would probably have already gone after her. In recent years their sizes had increased, as well as the volume of carrion they could consume. Working in tandem, he had seen three of them eviscerate a dead adult wandering albatross on its nest. Had they attacked it while it was still alive? Petrus did not know.

He tried to focus on the good things. But soon enough was tearing up with the thought of his mother. Osna, who would have been nearly 70 years, pre-adj, was the epitome of insouciance under naked pressure, never intending to stay on like she did, and then—without a choice in the matter—so fully embracing the career that made her famous among an elite group of, ultimately, nobodies. Such was the fate of those few victims left throughout Queen Maud Land. Victims of ignorance, for there was no way to grasp what had happened, actually, and where everybody else had gone.

Petrus now concluded that the entire weight of the continent, at least in remaining human terms, was on his shoulders. It was a considerable legacy. But in terms of an enduring chance, he was as fragile as a snowflake.

References

1. See "Males migrate farther than females in a differential migrant: an examination of the fasting endurance hypothesis," Elizabeth A. Gow[†] and Karen L. Wiebe, R Soc Open Sci. 2014 Dec; 1(4): 140346. Published online 2014 Dec 17. doi: 10.1098/rsos.140346, PMCID: PMC4448777, PMID: 26064574. Royal Society Open Science, https://www.ncbi.nlm.nih.gov/pmc/articles/PMC4448777/
2. See "This Catastrophic Polar Journal Resulted in one of the Best Adventure BooksEver Written," by Kat Eschner, Smithsonian Magazine, January 2, 2017, https://www.smithsonianmag.com/smart-news/inside-worst-journey-world-180961589/

Chapter 14
Quantum Ecodynamic Substrates

She had been as affectionate to her son as she had been dedicated to her craft. And, were it not for all that happened during Osna's life, might have made some enormous difference to the world. Molecular biology had certainly come of age. Her particular understanding of what stress did to amino acids, proteins, and phenotypic expression, had witnessed a veritable scientific renaissance. Epigenetic causes and consequences were all the rage. Specialists and funding agencies had come to believe in what she was doing on mountains, and along ice-free ridges; or in scuba gear within Antarctic lakes and pools, sampling phytoplankton transformations under varying conditions. Her goal had been to apply new chemical engineering tools in order to direct the evolution of certain random mutations which she knew could be rendered more and more efficacious. An efficacy guided by a desire to enliven *perfect* characteristics. Her admiration for James Baldwin's neo-Lamarckian insights into selection, and those heterophenomenological qualia of evolutionary philosopher Daniel Clement Dennett III [1], seemed to solve the *many minds* problem, even at the bacterial and viral levels.

Petrus, while revering his mother's prodigious skill, never stopped questioning the goals: what constituted *perfect*? And who was qualified to judge such things, when evolution had been functioning quite precisely for billions of years, engendering perfect beings, only to destroy them, quadrillions of individuals, including microbes, invertebrates and vertebrates, every day? To what end? An eternity of killing? He always asked the child's questions. What was the actual point of life? His former friends, and his mother, were always too busy to address that one.

Osna's lab at pes was sufficiently sophisticated to enable her to undertake gene editing, shuffling, in vivo insertions, and deletions. She knew that her husband didn't care, and her one son was never convinced. But she was. By applying the DSC (diversification-selection-amplification) end-products (protein variants) to various ecological stressors, like shifting humidity patterns, or signal-to-noise interference, Osna had demonstrated the first generation of quantum ecodynamic substrates. They provided, in her mind, undeniable evidence of an accelerated evolutionary process independent of what, in physics was known as counter-diabatic driving technology [2].

Petrus had understudied from her since childhood, obtaining the kinds of data sets, cured in every conceivable hardship, that tempered scientific implausibility with the possible. He had become, or was born, an innate optimist. But he rightly feared that the library of microflora his mother had collected and manipulated at pes was vulnerable. As fragile as the dozens of the bryozoans and other new species found 198 meters beneath the Ekström Ice Shelf back in the early 2020s. Pure invasiveness into entirely new ecosystems.

But he was essentially shut out of such ethical conversations. Those companions at pes might idly chat up such qualms, but their empirical training had left them bereft of theoretical humanity.

Moreover, Osna was defiant. No other facility, she argued, contained such proof positive samples of manipulated molecular evolution. They all objectively targeted species and genotypes—moss, penguins, glass sponges, tatters of 5500-year-old worms, fragments of bivalves. Osna had computed the probabilities. They all should be capable, once modified, of enduring draconian climate change and sustained solar emissions. They might yet combat the vast rise in cancers in the wake of sub-atomic particles, particularly the mysterious one, flaring from the Sun's surface. No one understood them. No one had yet resolved as to what precise particles were involved, or why the solar cycles were no longer predictable.

If only she could have communicated with the one man—her brother-in-law—who might have known, or at least known who to go to, so as to resolve the profound elixir, as she thought of it.

Medically, the Antarctic community had been in the dark for decades, baffled by the etiology, prevalence, and lethal levels of the particle impacts, not only on human health, but plants and animals throughout the in situ biotic communities. It was a mess compounded by the damned implacable silence from the outside. For all their combined genius at pes, they were basically helpless, feckless, useless.

Petrus stared at his mother at night. Sleeping on the floating island. There she was, Osna, who believed that she had discovered a radical group of amino acids, and specific evidence of a new level of protein folding that boded of an unprecedented *type* of immortal cell line [3]. She had no colleagues in the Antarctic, other than her son, who had any clue as to what she was really up to, or where it might lead. Certainties were less interesting than doubts. Causes more philosophical than effects. Mutations full of more possibilities than statistical norms. But human nature still preferred additions to subtractions. And it was still more metabolically tiring to look an organism in the eye, than to look away. Where these editorial decisions might lead a generation of biologists born long after the advent of CRISPR-Cas9 and other essential gene editing technologies was anybody's guess. Merging the cell lines of certain species of jelly fish with the last few remaining bonobos or Hawaiian crows? Or a sick child?

He quite precisely remembered the day the first four polar bears had been lowered by helicopter in a titanium net onto the ice, amid 60 mi gusts and after an immense journey from northern Greenland. Osna had injected them in their necks: a novel non-pyrogenic hybridized amino acid with electrolytes. The hope was that they might have more time to evolve in the great south than the rapidly fragmenting north. It was a good idea, in theory; like the orangutan translocations to Brunei from Kalimantan, along with newly gene edited rambutan (*Nephelium lappaceum*), their favorite fruit. But there was never to be closure, as far as anyone knew. The bears were never again seen.

Now, as he floated toward a sub-set of infinity, he knew that whatever his mother had discovered, none of it mattered. Lost presumably forever in the burnt out remains of her laboratory. And for that Petrus was rather relieved.

He embraced his fitting end, knowing what it must likely be. No national stage. But the absolute pleasure of total obscurity.

References

1. Beardsley, T. (1996) Profile: Daniel C. Dennett – Dennett's Dangerous Idea, Scientific American, **274**(2), 34–35.
2. Iram, S., Dolson, E., Chiel, J. *et al.* Controlling the speed and trajectory of evolution with counterdiabatic driving. *Nat. Phys.* **17,** 135–142 (2021). doi:https://doi.org/10.1038/s41567-020-0989-3
3. Salem, Mohamed (2020). "Design, synthesis, biological evaluation and molecular modeling study of new thieno[2,3-d]pyrimidines with anti-proliferative activity on pancreatic cancer cell lines". *Bioorganic Chemistry.* **94**: 103472. doi:https://doi.org/10.1016/j.bioorg.2019.103472. PMID 31813475 – via Elsevier Science Direct.\
4. Jacobs, R.D., and Samuels, P.Z., "National Genomic Academy Members Protest Immortal Cell Line Innovations," South Korean Journal Of Bioinformatics, Vol. 3, 2028, pp. 12-17.

Chapter 15
Looming Equations for Lost and Found

Within a matter of a few months it was overpoweringly clear to Petrus, and to the thinning penguins, that it was a different world than ever before. There was a sunrise and a sunset, by rapid successions; harbingers of spring, not the fall equinox, spring in the Antarctic. By Petrus' calculations this meant they had traveled not only northerly, but probably six, seven hundred miles, maybe more.

Runnels of water where surficial cracks had been gave credence to what was also quite obvious. The temperatures were soaring into the mid-twenties °C. He had ventured repeatedly close to the water's edge and seen Southern Right Whales, with their wind-catching flukes and white masses of whale lice and, specifically, an intense curiosity which kept them moving alongside the island. They could be migrating to Australia, or Brazil or South Africa, but it was always northward, seeking warmer water. Petrus had been told that the whales normally acquitted their northerly itineraries during the early part of October. But by now he was utterly confused by the calendars. Whatever timepiece had once been ordained, was now, as he thought of it, *obliterated*.

But this was no voyage through newly released heaven, as he also began to smell and notice the fetid, seething graveyards. These were dead zones of near zero oxygen (hypoxic), cyanobacterial cesspits. They were dozens of miles wide. Coming in the energy of upwelling, along with crucial but vulnerable organisms, from nearly 2000 fathoms, Petrus understood. Right through the eastern fringe of the Weddell Gyre, hitting the polar front deep, circumnavigating the open water. Dragging in their depths was a myriad of lost souls, from five species of krill and jellyfish never seen before, to every imaginable trapped bird, bioluminescent ghost cephalopods, humpback blackdevils, enormous ice- and toothfish and marine mammal. The vast majority of the by-kill of the deep mining excesses were species, whole genera new to science.

There was some movement still. The odd leopard seal struggling amid exoskeletons of cobalt and manganese chunks. Where there had been colossal wreckage strewn over hundreds of miles, repelling, and amalgamating whatever quagmires of

M. C. Tobias, *The Maiden Voyage of Petrus van Stijn*,
https://doi.org/10.1007/978-3-030-97683-5_15

former mining operations remained. Metallic dung heaps slammed into by icebergs the size of Hopi mesas traveling 40 miles per day. Ill-conceived drilling platforms guided by robots from the former satellite data that was dispatched from underground computer war rooms in Toronto, Santiago, or Shanghai to the inaccessible extraction sights thousands of miles away, out of reach of all protest or government oversite. The toxic debris fields had been jettisoned randomly over the past century by sudden top layer ocean shifts in the velocity of currents, thermohaline circulations long exacerbated by the intricacies of climate change, and then the total crash of the satellites.

The once ecologically impoverished microstate of Nauru had triggered this corporate frenzy to destroy the earth's last frontier, the deep oceans, in 2021 by pressuring the International Seabed Authority in Kingston, Jamaica on the basis of the inane and paradoxical claim that these companies involved were simply trying to ensure a steady supply of rare metals for the manufacture of batteries for billions of ultimately ill-conceived electric cars, regardless of the profound ecological consequences to the biodiverse-rich ocean floors [1].

There was no drag on Petrus' global needle, the old-fashioned compass indicating that he was passing between global zones 5 and 4. Magnetic variations had been accounted for, to the extent possible, and the declination was well west of what should have been a magnetic true north. But there were potentially a hundred items throughout the ruins of pes that might be altering north. There was no truth, no baseline by which an accurate directional might be ascertained.

But 1 day Petrus observed what could only be frigatebirds enjoying the transversal soaring in Southern Austral high winds and then, pointing downward his binoculars from the water's steepening edge, penguins breaching in the mighty swells, literally diving and flying between 120-foot swells. He had never seen such penguins before. But he knew from his studies that these were *Spheniscus demersus*, the South African penguin. His traveling companions, most notably the emperors, were one by one joining this odd species, diving into the roaring maroon walls of water. The impact upon Petrus was one of utter hope, but also a creeping terror. He sat up on his bunk bed, fighting sea sickness which told him that his island was rapidly shrinking. He was now feeling the full force of the roaring 40s, his archipelago of ice beneath him moving and sinking unrelentingly towards a new world.

Reference

1. As once described in a superb analysis by a journalist named Jonathan Watts. See Watts' essay, "Race to the bottom: the disastrous, blindfolded rush to mine the deep sea," The Guardian, 27 Sep 2021, https://www.theguardian.com/environment/2021/sep/27/race-to-the-bottom-the-disastrous-blindfolded-rush-to-mine-the-deep-sea

Chapter 16
Gardens in the Fog

Petrus could not know that he was reversing a trip by one Bartolomeu Dias, commander of the João II which had headed due south from Portugal in 1487, nearly 700 years before, all the way down the West Coast of Africa where Dias discovered Cabo Tormentoso, the Cape of Good Hope. Thirteen years later, Dias would vanish with his crew somewhere in the treacherous waters of the southern hemisphere, en route to India. According to historian Raymond John Howgego, the explorer left a stone pillar (padrão) somewhere on the Cape but otherwise virtually not a written word that remains as to what he actually saw there. He had been searching for the paradise realm of Prester John, allegedly adjacent to the African coast [1].

No mention of the graybeards, the legendary 200+ feet high waves of the Agulhas Current just West of the actual Cape. But presently, Petrus was awash with the phenomenal excitement of all that history, which had become the future. The entranceway through skyscraper breakers and burrowing crests. Of fog and a chaos of verdant horizons, scattered in and out so many miles just North and looming with an urgency that crashed through even the most steely of nerves. Such was the bewildering epiphanic, after weeks of dissembling pressure to all sides of his traveling island. The ice was pouring off the side as waves grew more and more ferocious.

Suddenly, for the first time in his life he was seeing what were unmistakably *trees*, *shrubs*, a ground cover higher than the highest of the tall grasses of his native home.

The island was being blitzed by thousands of whooping seabirds, their excitement breaking out. Sun seared the moving leviathan of dissolving ice. The foaming loft of giant azure pinnacles crashing with unimpeded ferocity, this was all new to Petrus. And the temperatures continuing to warm.

A group of three penguins had washed up on to the ice in one of the towering swells, condemned to a stinking eddy of garbage. Golf balls, pieces of kitchenware, tarfiles and a trove of bottles, wrappers, human castoffs from every corner of the planet. The three avians were trapped together in metallic netting and had been for

M. C. Tobias, *The Maiden Voyage of Petrus van Stijn*,
https://doi.org/10.1007/978-3-030-97683-5_16

some time. Two were dead, the other dry heaving. When Petrus tried to save her, she regurgitated a slime thick with the fragments of a surgical glove. Then her eyes closed forever. Petrus threw all three, freed of the mesh, back out over the ice. He was benumbed.

More swells brought more insidious trappings of what had become of the west African main.

The last of the emperors had left weeks before, all emaciated, heading to dreamt of places. But one, whom Petrus had always thought of as Anatole, had hesitated to desert his friend, the human, of so many years. In the end the heat was too much and the poor bird, like so many of the penguins, slipped away into the forever depths with a muted farewell that cannot be described.

Days later, Petrus stood shakily on his experimental surface, ankle deep in melted ice. But for all that had happened, excitement gripped the nervous lad. Every second now, as he seemed to be heading straight for the Cape. Then, as suddenly, the mass of ice was distinctly being jettisoned more northwesterly. He would miss the continent, instead moving at a visibly calculable pace somewhere up and up forever upwards, it seemed. For weeks, the adrenalin overpowered his otherwise stable composure. He knew, of course, that he must have traveled a few thousand miles and those sums they kept adding up as his raw journey into neverlands continued day and night. Through rainstorms the likes of which he had never witnessed. Rain that soaked his mouth and he threw himself into it with a startled sense of re-birth. Scooping out water from the remains of his bedroom. Never, never had water seemed so perfect, to one who had escaped the driest, highest continent on earth.

A warm fog now settled in upon his shipwreck of ice. It was enraptured by encircling rainbows. An enchantment that crowned every corner of Petrus' daylight hours, giving vent to constellations by nightfall whose clarity amid gaseous infinities of the Milky Way now mentored him in all the important lessons of drama. He could hardly breathe, such was his exhilaration. He just kept breathing.

So it went, this way, with humpbacks off the warm waters of what was actually the coastline of Ghana, breaching fantastically. Schools of Atlantic Humpback Dolphins constantly at play. While albatross and other seabirds entirely new to Petrus swooped across every brow of the icefloe, having never probably seen such a fortress of unpigmented, behemothic momentum. The white virginal castle was now far up the coast and coming near to the southern Mauritanian hotlands, where Guinea's verdant crags gave way to a dry valley-like desert of ribcage monoliths, rocky, Saharan, with crazy dunes that were of sand, not ice. Petrus was glued to his binoculars as the Sun continued to burn down more and more fiercely. The rate of ice melt could not be sustained. At some point, mathematics dictated an absolute irreversibility. He was, by his sufficient training in even casual triangulations, not 30 mi from the shorelines passing by. They were visible with each enormous rising swell that threatened, from all sides, to overwhelm his much-disintegrated iceberg.

But this sensation was confounded by the fact ice was amassing from the north, drifting down in swales and bergs, in flat melting rills, dells, and cascades of what could only be Arctic calving. Never to his knowledge had ice from the two poles met like this. He knew it had to be consistent with the long-time shutting off the

Atlantic circulation forces. This incoming ice was slightly more pale and pink in color, a sense of blood, blood and salt with nowhere to take leave but downwards, dissolving any solids in its unrelenting course.

Sensing by all his 30 odd years of uniquely skilled Antarctic acumen, he outfitted the remaining parts that together could reasonably be called a Zodiac, with its peculiar form of power. And to be certain, placed two unburnt paddles in its hull. He suited up. Time, too, was fleeing the remains of the ice. He was soon to be engulfed. And wet through and through.

Reference

1. See *Encyclopedia of Exploration to 1800*, by Raymond John Howgego, Hordern House, Sydney, Australia, 2011, pp. 308-310.

Chapter 17
The Legacy of Thomas Ralph Merton

He awoke to the startling appearance of a signal within that vast arena of background noise that was, apparently, the opening to the Mediterranean near Tangier [1]. There, at dusk he spotted faintly what could only be the legendary camel, many of them on a high ridgeline slowly plodding through a dust storm with the last of the sunlight piercing their hallucinatory trail. *Camels!* Petrus thought. *The Arabian nights*!

He was beyond himself with elation, as he watched from just outside the sinking metallic melodrama of the final portions of pes.

When just as suddenly, darkness now aboard, he heard the unimaginable: shouting from across the choppy waters. Far beyond, circling under the stars, the moon illuminating it, stood a huge lone tower. Down below was a large incoming dinghy with four men paddling out from rocks and the huge crashing waves, and from what (he would soon learn) was called the Europa Point Lighthouse, dark for half-century.

There were only moments to spare before he made ready to set off into the frigorific but friendly men's worn old skiff, as a discernible face to the shouting voice rose over the whitecaps. At that moment the last of the teetering chunks of ice crashed in upon his once archipelago. The ungainly deracinated ragamuffin grabbed an extended arm and leapt into the rescue vessel.

"And where have you come from?" a haggard old fisherman-like voice hazarded. Three elderly but brawny cohorts deftly relocated the voyager, as the last remnants of Petrus' frozen world drifted downward, like a cloud-teased whisp by whisper into a leaden oblivion.

The empire of pes, with a century of data locked in waterlogged computers, now lay more than 450 fathoms below, where the Mediterranean met the Atlantic.

© The Author(s), under exclusive license to Springer Nature Switzerland AG 2022
M. C. Tobias, *The Maiden Voyage of Petrus van Stijn*,
https://doi.org/10.1007/978-3-030-97683-5_17

Reference

1. 1888-1969, Merton – a true Renaissance man - had the last physics laboratory in all the UK that was still in private hands, his. Rich, polymathic, he was in love with light, particularly the effects of diffraction grating. His brain was a prism.

Chapter 18
Petrus Discovers Wonderland

Even to blink was to risk melting the invisible, disturbing all the real miniature horses and large walking swans along lagoon banks festooned in floating gardens, the odd gypsy drying a seventeenth century rug inside courtyards whose Morish pavers and curving, painted architecture had not changed since the height of the Renaissance. Every object, gesture, nuance hinted at, was grounded in the antiquity of a hallucination, illuminated by the odd campfire here and there. Tens-of-thousands of diverse birds looked on in the dark, eyes gleaming, as well as Amazon like moths, swarming in the warm night. No such thing as a tourist in 60 years, adj or pre-adj. Total population of Gibraltar estimated at 19, his rescuers were quick to point out.

"Gibraltar?" Petrus exclaimed. Just saying the name spun wild fantasies in his head.

The medieval and renaissance town had been so abandoned by 99+ percent of the deceased population as to have fostered a veritable rewilding of a city into a foremost Utopia like harbor. It was dominated by the fourteenth century Tower of Homage and old wall, upon which a lengthy row of lanterns were lighting his way, as he was escorted to the townhall, where there was a bed he could sleep in, and a waiting contingent of enthralled bureaucrats.

Or, for that matter, there were countless other empty beds throughout the village, the choice was his, said an elder.

"I'm quite at your disposal." Petrus, at ease in the role of a Captain Cook or Lemuel Gulliver. He knew those stories.

Once he was settled in at the townhall, a group of perhaps ten locals who had arrived to greet him, conveyed their bundles of welcoming bouquets and fruits to devour at his pleasure. They were, it appeared to him, farmers, and the astonishment on their faces was keen. They carried torches. Wide-eyed and curious, no one dared to be the first to ask a question. There were so many half-baked ideas skipping through the muggy ambience. A silly contagion of first thoughts, Aristotelian primaries that evolution has dictated at that primordial intersection of strangers in the night. Petrus himself was no less bewildered.

© The Author(s), under exclusive license to Springer Nature Switzerland AG 2022

M. C. Tobias, *The Maiden Voyage of Petrus van Stijn*,

https://doi.org/10.1007/978-3-030-97683-5_18

If he had arrived from Tangier, the first to do so in decades, how was it he arrived on an iceberg? This was the utterly insurmountable concept that was sweeping through the whole community of Gibraltar, seen and/or unseen.

But then, Petrus was equally famished with respect to information, and not a little queasy after however many months at sea. He was feeling a sensation not una-kin to vertigo, the stabilizers in his cranium doing double takes. Standing, walking on solid ground.

A dog much like Max came up to him to sniff out this most peculiar man. He did not growl, but at once accepted him as if he were a welcome offering.

"Don't worry," said the dog's human companion. "He's as loving as a fly."

"A Congress Pear," said another.

"A welcoming pillow," someone else volunteered, vying for the most apt description.

Finally, the astonishing business of Petrus' visitation was raised. The trepidation of the locals was no less than some official expression of Marian dogma, or Anastasis. "What are you doing? Where were you going?" And the lingering group suspicion that they had all missed out on something. That, or the ever-possible arrival of the real Messiah. The interest showed was heavily shared. All eyes were glued and widened.

Petrus tried to explain in as few words as possible, which accorded with the style by which he was raised. His mother had been a stickler for concision. It was a habit born of freezing air and the dangers of harming one's lungs. "You only have so many words before it becomes dangerous to utter them," she had taught him, loosely after Jonathan Swift.

"Belgium. I need to get to Belgium," he replied to the group of locals thirsting after news.

A debate ensued. Was it possible? Of course, it's possible, went the consensus, though Petrus discerned the unmistakable alarm in the general sense, as the place-name, Belgium, really seemed to sink in.

Soon, an elder among them volunteered, "It's too complicated in the dark. He's tired, can't you see? Let the boy sleep."

"I'm actually going on thirty-one," Petrus said, shyly. He had not been called a "boy" ever. And noticed that everyone around him looked quite old, in fact. More women than men.

"What is this?" Petrus asked. He had been handed a pomegranate.

"Umm," he said, tasting it. "Remarkable. And what is that?"

"Calentita," declared an elderly woman, her growth clearly having been stunted along the way somewhere in her life, in a dun-colored apron. She eagerly enquired, "¿te gusta?"

He tried it. It was still warm, fresh out of an oven. "Delicious."

Finally, "He needs his rest!" another man reiterated anxiously.

They could not do enough for this combined callow, yet Sir Edmund Hillary-like veteran. A shocking admixture of fascination to everybody's system. They showed Petrus the way to a real bathroom, with a real shower, and all the hot water he could possibly waste, heated from an oven behind. A little kitchen still proffered the old induction hobs.

Chapter 19
In the Morning

Petrus was awakened not only by the crowd of locals whispering audibly, but a colloquy of curious Barbary macaques, one of whom jumped onto Petrus' shoulders.

"That's Cactus," said one of the locals in Spanish, laughing. "He can be feisty."

"I don't speak your language?" a diffident Petrus explained. "I'm sorry."

Someone translated.

"You have so many monkeys here," Petrus replied. "They're amazing!" He was stroking Cactus gingerly.

"Rock apes," the man went on. "We are told they have spread across Europe. Always friendly."

Petrus was mystified.

People shrugged and seemed amused.

During a brief tour of the village, it was clear to his escorts that the boy was more than merely peculiar, certainly no threat. Everything excited him, appeared novel. His innocence spread like a contagion, received by one and all gratefully. After all, they had hoisted their suspicions by dint of their very crowd gathering mass around a potential invader.

"Have you been in a cage all your life?" one asked rhetorically, smiling at the thought of this strange young man who stared around in wonder, as if a row of thatched domiciles, an old church, wild horses that were not, the occasional giraffe, Persian cat, dragonflies converging over the marsh were so out of the ordinary.

"This is my first time in Europe," Petrus imparted. He explained where his voyage had originated when asked about the iceberg from which he was rescued.

"No?" conjectured one of the rescuers who walked with his wife. "How can that be?"

Both were dressed in cloth boots, cloth belts. She was wearing a poneva skirt and blouse decorated in a simple brown onion color patterned in a few humdrum medallions and stripes.

Everyone wore large sun-protective hats and wrap-around model dark glasses, the reason for which was obvious. Petrus wore his own, of a nickel alloy always

© The Author(s), under exclusive license to Springer Nature Switzerland AG 2022

M. C. Tobias, *The Maiden Voyage of Petrus van Stijn*,

https://doi.org/10.1007/978-3-030-97683-5_19

prized, which had survived the catastrophes at pes. The locals appeared uniformly sunburnt, to various degrees.

Petrus was not willing to press too hard on any one topic, though so many were swarming in his loft. But he was hesitant to risk insulting or even conveying the slightest nuance of the differential, as his mind calculated risk factors amongst a crowd of odd, though friendly fellow humans. He feared any possibility of an altercation. He had seen one or two among station personnel over the years, especially when the solar flares had worsened, affecting every aspect of personality and mental health. Here he had every reason to suspect a similar possibility, if not more so. The prospect of contretemps had never burdened Petrus, however. His reading of natural history had precluded dissention any more cardinal than that of a Serengeti, or a pod of killer whales manipulating a doomed seal off its transient refuge of an iceberg. He was prepared and could chop a log or hurl a heavy stone like anyone else. He possessed a fitness to have survived perhaps one of the most arduous upbringings of any young man in history.

Admittedly, he had never witnessed a demonstration of anger amongst a penguin rookery, only helpless indifference. But the equation for large, compacted numbers of mammals was an entirely unknown entity to him. He imagined that no good idea could survive a mob.

"There are camels in Gibraltar," Petrus stated, as if establishing a fact for its first time. Every sensation was tied to importunate incredulity.

"Ever since I can remember," James, the chief tour guide, probably 80 years old, replied.

"And they're friendly?"

"Why wouldn't they be?" James' wife, who spoke a British English—Petrus certainly knew the accent—said, almost defensively.

"What do you think the temperature is?" Petrus asked, standing before a village grocery store. His thermometer had not escaped pes.

James shrugged. "In the low 20s, no doubt. It usually is. Gibraltar is said to have the mildest climate in all Europe. Although the summers are now becoming very uncomfortable."

"At least for the last few years," his wife, Beatrice added. "But we are of course cautious."

They still spoke in pre-adj and calculated in Celsius.

Now Petrus removed his pocket compass and checked the needle. It was shivering meaninglessly.

"May I?" And Petrus entered the store, studying contents on wooden shelves as if it were a priceless museum of colorful artifacts. Fruits and loaves of bread in abundance. A screwdriver. A wrench. Chocolates in home-sown little bags, with the word, Cacao printed on them.

Back outside, he looked to the high rock, and along all the ridges, searching for a radio tower.

"No radio towers?" he asked.

James shook his head and said, "Once, of course." He went on to explain what Petrus had feared, thinking that just possibly the moribund history of the twenty-first century might have gone differently on other continents. That just maybe, Queen Maud Land had been the one planetary anomaly. It was not.

"May I ask you a question?" Petrus entreated. "Where are the children?"

James glanced at his wife, then began, "You really don't know, do you?"

Petrus did not press, his puzzlement bespeaking volumetrics. Finally, Beatrice added, "That part was simply terrible."

Chapter 20
1400 Miles to Bruges

That evening Petrus looked over maps provisioned by James. It would take him approximately three months to reach his destination, assuming he sustained 20+ miles per day. It was an ambitious itinerary directly through Castile and La Mancha, then all across France. He looked forward to Paris. Indeed, to everything.

"But you have a distinct advantage over almost anyone you are likely to meet. They will be assuredly astonished and hospitable. Because you are unlikely to find any as young as yourself."

"And now is the season for French beans, courgettes, cherries, potatoes, white asparagus, all there for the picking," said Beatrice.

"I have no money to pay for anything," Petrus admitted.

James sighed. Beatrice smiled. "There is no money, not anymore," she said, explaining, "Everything is free."

Petrus, knowing absolutely nothing about economics nonetheless could appreciate the counter-intuitive arithmetic cul de sac in that logic.

"But surely there is supply and demand?" he asked. "How do you avoid capitalism, people hoarding?" Gibraltar was no scientific research station where a Utopic orientation was possible amongst those few resident collaborators. Where the finest technologies had been outfitted for the specific living conditions, at least until the end.

James deferred to Terrence McCoy, a long-retired college educator, with a glorious mop of robust white hair, a stalwart type, ponderous until this moment, and perhaps even older than James, who ventured to explain. "You see…" and went on to elucidate the de-population quotient, which maximized individual profit in so extreme a manner, and so rapidly, as to redeem poverty with egalitarianism. "Much like the famed inflation in Hungary in 1946 where prices for daily goods more than doubled every day," he added.

"I don't profess to even know what terms to describe a world economy that has lost all its engines of opportunity. There are no more labor forces, motivations to

M. C. Tobias, *The Maiden Voyage of Petrus van Stijn*,
https://doi.org/10.1007/978-3-030-97683-5_20

become greedy, and no consumer index. How could there be, in absence of nearly all consumers?"

He tried to convey concepts of meaning and meaninglessness in the new monetary reality. It was no system but a de facto free for all, without limitation. Somehow remarkably, the dearth of constraints did not break boundaries or friendships. The depauperate community had become immensely wealthy by dint of nothing more than freedom in its most perfect, absolute forms.

Deflation, multiplied by zero opportunism, meant that the social rules had all changed (notwithstanding the bizarre nature of multiplying anything times zero). As Petrus understood what Terrence was describing, there was simply no reason to worry about most things.

"Although as has always been the case throughout history," James added, "two professions have always prospered more than any others. That of doctor/pharmacologists, and bakers."

"They too work for free?" Petrus inquired, an insatiable appetite for as much news of this transformed reality as possible.

"Therein lies the basis for the existing barter system. Nothing primitive about it. Practical, fair, and efficient," explained another sprightly octogenarian standing beside James and Terrence and Beatrice.

"The problem," said James, "is finding a doctor when needed. Bakers are more plentiful. My wife, for example."

It all made vital sense to Petrus, who had known only a matrix of privation requiring not only years of rationing, but the predicative mathematics of a grim future wreathed in disassembly, something far more grand that mere chaos.

That night his sleep was both dazzled and dismayed. He thought of Osna's final moments, and the regrets he knew he would always carry, for not being able to bury her. She had shown him everything yet must have kept so much of her hurt from ever turning over. Her stubbornness was, one could rightly say, her undoing, in the sense that she never made it home. She never saw her husband again, or her college friends, her parents, all the places that had made her who she was. That inward repudiation had caught up with her early, as it did all the occupants of pes, each harboring their own self-accusations, long before the actual end.

Petrus went to bed each night with his mother's doom planted in his open eyes.

But he also recalled the eyes of Anatole gazing upon the tall companion. The waning light in those orbs which clearly anticipated all that would benumb the bird, whose lifespan might ordinarily have been 15, even 20 years. These facts of life were heartbreaking and only a philosophy that clearly acknowledged the futility of too much grief could hope to survive such complicated events in one's past.

In the morning Petrus graciously thanked his hosts and bid farewell. Walking asymmetrically northward, stopping it seemed, every few feet to gaze upon altogether new life forms.

Chapter 21
Adventure in Extremadura

Many days had passed, in wonderment.

Just before dark, Petrus settled into a cosseted bower, a former extension of a baked brick mill, circa Renaissance, he imagined. It adjoined the ruins of what was clearly an ancient outdoor theater. He slept within the bivouac sack given to him by James, and a nicely compressible down pillow Beatrice had fetched from her own bedroom. As he prepared for the dark, a strange group of creatures standing upwards, perhaps each a foot and a half to two feet in height, came from a nearby forest, and gathered round him with questions they asked in an animal language Petrus could not understand, of course. He knew he recognized them from some biology book or other but could not recall even the common name of their species. Perhaps they were a new sub-species of meerkat. Golden. But they exhibited no canines in their smiles. They were as friendly as could be and appeared to be impressively well-groomed.

The previous week had left Petrus more perplexed than ever. Not just the novel biological profusions everywhere he cast his glance, but on account that, once having departed his hosts in Gibraltar, he had seen all of four people, farmers, each speaking different languages or dialects when approached. All of them appeared to be spooked by the sight of the newcomer. So unsettled, in fact, that they trotted off, inexplicably fearful. Their dogs or horses, goats, or sheep would prance away with them, silently in deeply personal silos of superstition.

Of course, each of these mammals was to Petrus' eye, the most spellbinding of creatures. Even the dogs were different from those he was accustomed to back home, on the ice. But what most astonished him was the forest and its birds, insects, and so many others. Parrots, gold and deep cerulean. What a remarkable group of birds, he understood at once, whereas previously, they had just been beautifully painted things, on paper. And the oak trees, hundreds of feet high, probably twenty-person girths, festooned with co-symbiotic life forms. Every tree was an Amazon all its own.

© The Author(s), under exclusive license to Springer Nature Switzerland AG 2022
M. C. Tobias, *The Maiden Voyage of Petrus van Stijn*,
https://doi.org/10.1007/978-3-030-97683-5_21

After a night listening to the howls all around him—hyenas, he had to assume, and owls hooting—he awoke to a frightening, but not surprising sunrise. It was blinding. Not like on the ice, where there was no precise accounting for the Sun's intensity, layered in fogs, but a grey gold, surrounded by moisture-laden umbral ellipses, cirrostratus strips of cloud fast moving. He at once donned his dark glasses. They would become his most precious item to safeguard.

He stuffed his rucksack and moved on. Every mile was a new adventure, new breath, new sky, new horizon. A cavalcade of unknown nomenclatures. Striking forms of Corona Pine, even of Eucalypts. Absolutely everything landed in his eye like a bestselling romance novel, and no sensation could compare. Imperial eagles, little kestrels, bustards, and sparrow hawks. Nightingales and black storks. Of the plants he easily recognized various genera—Tulipa, Euphorbia, Dactyloctenium, Cyperus. Enormous fresh paw prints in sullied mud, unmistakably elephants, or so he assumed. Every occasion called for a new incredulity as he endeavored to re-write any previously assumed norms of biological dispersion. The geography did not conform to known territories of life.

One morning, 9:15 by his diving watch, he arrived upon an elegant ensemble of ruins near to a hamlet. Before him stood the four roofless walls of a Roman-looking church. Of course, what did he know? Snow-capped peaks beyond. Snow which, even at that distance Petrus could easily divine. It was the roseate color of slush. Atop one of the crumbling walls rested a giant nest. Almost at once a stork landed to feed its young. Adjoining the ruins, there was a woman, not running off but, rather, approaching him with the hint of a smile. She wore a long skirt, brightly decorated. A girl, more precisely, probably no more than twenty.

At a comfortable distance she there stood her ground, a sensuous study in curiosity. Some sheep scooted behind her.

"Hello," Petrus beckoned.

"Hi," the resplendent young woman replied softly shyly, mumbling something to her sheep who was nudging her affectionately, "*José, deja de masticar mi vestido!*"

"Where am I?" Petrus asked.

"Where you are?" she repeated.

"Yes."

"Near Cáceres?" she said with a teasing question mark.

"Cáceres?"

"Yes, and there," she pointed up and away, "Sierra de la Mosca. This is Extremadura. You don't know?"

"No." And then, "What is your name?" he said. "I am Petrus."

"Petrus." She stood forward to shake his hand, gazing at him. "My name Dulce Roso," she volunteered proudly.

"You speak English, then?"

"A little." And she repeated, *Petrus.*

He nodded. "Perfect. Petrus."

"We see little time strangers," she went on to explain. "Really almost nobody. So, Petrus, where you coming from?"

"Gibraltar," said he, to make it simple.

She motioned for him to come along with her.

He followed her to the small grouping of houses, just beyond a low ridge where the vegetation benefitted from a spring. Three farming huts surrounded by a deep sylvan glen. Outside one of them, Dulce's own, was her pet lynx, whom she called Margarita.

"So beautiful! She doesn't chase the little birds?" Petrus noticed how unconcerned all the magpies, crows, and sparrows were feeding on the ground.

"You're funny," she said.

"Okay. So what does she eat?"

"She likes lemon cupcakes, bowls of cereal, like that. She even drinks Margaritas, of course."

"Come on."

"Really. Why not?"

Now two of her neighbors had come to see what the excitement was about. And with them, two enormous hornbills, and a New Guinea flightless rail that ran, hopped to keep up with their companion humans. One of them had a fixed patch on its left eye.

This is all crazy, Petrus thought. Margarita did not stir. Snugly fitted at the entrance to Dulce's front door, fashioned of old, carved quercus.

One hospitality led to another and by nightfall, Petrus found himself alone with Dulce in her two-room pequeño alojar. José had come in near to the mellow hearth with its licking flames drawing warmth from a redolent pine to consume her dinner, then left abruptly. In Margarita's dark pools of eyes, Petrus saw the eternal dream of flickering light that bode of all things good and fair. Of a life determined to be free, casual, not overly interested nor disinterested. Just right. Like the Jain Digambara Sādhus he'd once read about.

"Your husband?" Petrus asked, trying not to stare at the beauty before him, with her piercing green eyes and lengthy black braided hair.

"No," Dulce said matter-of-factly.

"I'm sorry."

"No husband," she repeated. "Bad luck behind me."

"I see."

That night they stayed in the same bed. Petrus had never been with a woman. She undressed his stinky clothes. Her hands were the color of tea leaves, prayer beads, chapped lips, of endlessly exchanged coins.

"You have been on the road a long time," she posited.

"I have."

She noticed a scar on his shoulder and neck (where he'd slammed into the steering column of the hydroski).

They touched one another, as if disarming dynamite, so gently, welcoming a new kingdom, fingers treading with the forgotten art of touch over a lost treasure map of original flesh. Petrus bid farewell to his virginity amid a quivering shock to his long-dulled system, though if you were to ask him, nothing whatsoever was lost. She was to become an ambrosial memory, one he could not extirpate. He thought to ask her if she would like to journey with him.

"I could never leave," she admitted. But it was clear to Petrus that the idea of leaving presented an insurmountably frightening prospect.

"Surely your neighbors would take care of your pets?"

"They are old. And they are all dying."

"What do you mean?"

"We are all dying."

"Of course, but you are young, the youngest person I have ever seen, in fact?"

Petrus tried to make light of the moment and pointed to his own head. "This will be a skull one of these days." But he could not make the grin last.

She wouldn't engage further in such nonsense. To her, it was mere baiting.

The next morning, she pointed out a beautiful cluster of flowers, gently twisting, playing with their own four-inch shadows in the breeze.

"It only lives a few days," Dulce remarked, curious to know of Petrus' thoughts.

Petrus knew the flower. But he thought the species had long ago gone extinct.

"You know it?"

"Ramosmania rodriguesii," he said.

Dulce giggled at the ridiculous sound. "What kind of name is that? How do you know?"

"I just do. Also, Café marron," he added. "Common name. It cannot fertilize itself and is, as a result, said to be very lonely."

She stood back. "It doesn't look lonely. There are several of them there. Anyway, I like any good Café. If one only existed. But, maybe a boy or girl flower finds its way here and what do you know, this one becomes not so lonely, eh? Of course, I'm just teasing you Mr. Botanist."

"Maybe so," Petrus configured in his head aloud, a smile to re-assure her. "Where did you learn your English?"

"My father she said," putting the subject to rest.

A burro easily pushed open the heavy oak front door, greatly accelerating Petrus' enthrallment.

"That's Hiccups," Dulce said, giggling. The 500-pound equine came and settled down beside the bed, wanting to meet the stranger. He lay upon broken red pavers upon where the bright morning light shone in the aspect of palimpsest. And then a second one entered, also making herself at home.

"Iris," Dulce added. "They're inseparable. Iris is blind and would be lost without her companion."

In subsequent nights, far removed from the ruins of the old church, and the cloud-swept mountains, he would think of her; and that tremendously rare flower and the two donkeys which apparently never cultivated a taste for those explicit flowers, as inhabiting their own Eden. But then, every vortex to which his itinerary brought him, also veered towards a similar, perfect image, unsullied, somehow adroitly preserved in the unfailing *memory* even before the experience of it. The future past, or metaphysics, he reckoned.

"Iris, José, and Hiccups," © By M.C. Tobias

Chapter 22
The French

He reached France within a week. Petrus spoke their language. He did not expect the size of the ungainly Eiffel Tower, portions of which had been blasted out, desecrated, including the upper most portions, where tourists once flocked. It had been severed entire. Nor was he prepared in any way for the solemn experience of Notre Dame. It was late morning.

"Bonjour" the voice carried, not as an echo, but empty vessel hollowed from within. There were all of two other people in that enormous ruin, as far as he could decipher. There was a very ancient woman sitting in a pew stroking her mangled cat. Over there, beneath the haunted west rose window, modeling the auras throughout, was a man doing something. Petrus moved closer. He was performing some kind of surgery.

"Qu'est-ce que c'est?"

"You're Flemish," the man at once blurted. "Or from England?"

"Belgium."

"Well this an FI hybrid, a Mulard."

"A duck."

"Of course, a duck."

The animal lay tranquilized upon a surgical table, beneath the fantastic light of the early thirteenth century-stained glass above.

"What's wrong with him?" Petrus asked.

"Her. I'm injecting some oil of Mackerel. She's got serious cataracts and I would hate to see her go blind."

Otherwise, the man showed little interest in Petrus but upon questioning informed the traveler that the entire human population of Paris was slightly over 50 persons, or that was the common assumption.

"Fifty? You're sure?"

"Fairly certain. Much to the delight of everyone else, especially these guys. Now, finally, we French are unburdened."

© The Author(s), under exclusive license to Springer Nature Switzerland AG 2022
M. C. Tobias, *The Maiden Voyage of Petrus van Stijn*,
https://doi.org/10.1007/978-3-030-97683-5_22

"I don't understand."

"It doesn't matter. You're too young to get my meaning."

But Petrus actually did get and assumed he was referring to other animals that would now be spared the notorious traditional French diets. And he'd observed a flock of Eurasian Tree Sparrows somewhere south of Paris that had to have numbered a billion individuals or more. And pigeons and ducks everywhere, it seemed.

The river through town was no longer fully encased in the limestone buttresses. Many of the previous platforms had disintegrated. Wars, fire, scavengers, who knew. Most of the inner peripherique, he was informed by an old monk passing by somewhere north of Notre Dame, had long ago been abandoned. Much that had been built after the time of the civil prefect of Seine, Georges-Eugène Haussmann, two centuries before, had been dismembered, much like the former monastery of Cluny. A whole new hagiography of modules had been erected, typically in hurried fashion for temporary relief. A thirty-five-mile extension of the Bois de Boulogne now reinvigorated what had been the metropolis. All the so-called arrondissements had been disassembled back into a borderless wilderness throughout, where elephants, gazelles, gorillas, and other herbivores roamed freely.

This was a Paris that Petrus had never imagined, with vast geese gaggles waddling down otherwise empty boulevards. Coyotes napping under café parasols. An old man dressed like a clown juggling before a bored 3000-pound hippo who lay dreamily on a sidewalk, its baby beside her.

Petrus had plenty of time to contemplate the implications of this bonanza of biodiversity but was unclear in his mind how to reconcile it with all that was happening, in general, particularly to the avifauna of the Antarctic. Moreover, he had to get food, water, and stay alert. He'd kept to his pace, though one continually punctuated by pauses, as when he surprised a pack of hybrid wolves that were eating grass. To his amazed relief, they were all somehow incredibly docile, as if fully domesticated by humans, tails wagging. They wanted to play with Petrus. He was no fool: it made absolutely no sense. There had to be a hidden danger in all this.

He slept as a guest of the town that night in the royal chapel of Sainte-Chapelle, with its 6450 square feet of thirteenth century-stained glass depicting scenes from the Bible and commissioned by King Louis 1X, he was informed. A cozy little bed of Egyptian cotton, years before taken from a suite at the Hotel George V, served as his resting place. The winding stone stairs up into the main body of the church, constructed before 1230, gave birth to not so much a *new* Petrus as a profoundly perplexed one.

This was especially so after he was informed that no visitor had entered Paris in over a year, that anyone knew of. And certainly, none from as far away as Gibraltar. He kept other details of his much longer journey to himself, as there was enough information to exchange without causing the kind of apocalyptic scare that, he thought, might easily swing the balance from that of a welcomed guest, a downright curiosity, to the perception of some kind of pariah, a disease messenger. Credulity hovering upon suspicion, or superstition, was in the air. But it did not undermine the congenial hospitality shown the stranger wherever he wandered.

Late into the evening half-dozen elders of the Paris Commune stayed with Petrus, offering him a generous sampling of some fine local bread, figs in goat cheese blankets, olives farcies, Bleu d'Auvergne, and Camembert.

"Heavy on the cheese," Petrus stated plainly.

"It's not Fouquet's but then, alas… Here, try this, a freshly cooked thick vegetable pistou, and selection of our local wines." The elderly woman, Aminé, served him, and availing of the opportunity at hand, all asked him a rash of questions. What was the situation in Madrid? Were their indeed outlaws to the south, particularly near Toulouse? Was anyone, anywhere, in any kind of uniform? Had he seen *any* antenna, or heard a radio? All very relevant questions to which Petrus answered as best he could, then eventually raised the one topic nobody during his journey thus far had addressed: "What happened, where is everybody?"

How could he not know? They wondered.

"I was in a remote area of the Madeiras," he explained.

"And there are survivors on the islands?"

"Yes absolutely."

"How many?"

"I didn't count, but not many."

"And the Sun cancers?"

"Yes. Carcinomas are widespread."

"As you can rightly see, Paris lost over thirty million people," the very old Father, Bruno, declared, and gestured from Corinthians that all their souls may be saved.

"In what timeframe?" asked Petrus, "not to be cynical."

"Two, three generations," another who had been grotesquely blinded, said. "I was a lucky one."

"No Salvatore," Father Bruno gently offered. "You have suffered."

"He was struck, nearly slain," the Father added.

"And the children?" Petrus went on.

Those present looked to one another. Bruno, stoic and gentle, replied wearily, "Very few survived, especially in the beginning. They were the most susceptible to the Sun, you know that?"

"I was in a remote place."

"All your life?"

"Yes." And after an awkward pause, "And now?"

"Now? You ask as if you are magically immune, or simply new to everything." Petrus nodded dumbly.

"You should understand that we are all resigned to it, quite happily so, I might add. One or two children made it through the worst part of those times. Is it wrong to admit that, just maybe, these are indeed the best of times?" His friends all visibly concurred, not that anyone was necessarily thinking of Charles Dickens.

"I don't see how it can be construed differently. It is not wrong to volunteer such things, even before God. More of everything for everyone. Greed is gone. Much, much more without any quarrels. The old saying, that less is more, well, the world certainly upended many meanings of many things," Father Bruno elaborated, exploring for himself those avenues of penance that were open, shying clear of those that might be closed. "If there were sins, they have been more than received and absolved."

"Of course, nobody is certain," a detracting voice said hesitantly.

"The Church does not add up all the deaths and survivals, if I might suggest," intoned one of Bruno's junior officials in the sacred hall, Maxime, not 50 ft away from the Holy Fragments of the Crucifixion. "This is not mathematics, but the new life God has given to each of us, to make sense of. Not to calculate who did what to whom, who was right, who was wrong, as if the sum-total should translate into vengeful reciprocity or karma. Rather, to make peace with what has happened and what is next. So that, at the end, we don't look back at our lives and regret that all we did was criticize the world. A reasonable person might find the twenty-first century a terrible one into which to have been born. Too much to find fault with. To question whether there really is a masterplan, or simply perpetual torment, as so much of European history attests, every century filled with its tragedies. But also comforting small talk. So much love. Love that is never diminished the more it spreads. So, I'll stop now. The little details speak for themselves, wouldn't you agree?"

Bruno lightly bowed in a show of support.

"Yes," said Petrus. He liked these Frenchmen. But he was also computing in his head. The population of Paris, if his newfound companions were to be believed, numbered fewer than 0.000002 percent of its previous population. Extrapolated globally, though with only the roughest, unstudied schematics in his head of rural to urban distributions, and so forth, that would mean that the total human population might number as few as 25,000 to 50,000 souls. That represented an astonishing reversal, given that the last known census in the late 2020s exceeded ten billion. At that time, several demographic regions were still rapidly increasing in number.

Bruno produced from his cloak a small work of prayer, a highly coveted book of hours, he explained to Petrus. It was known as the *Hours of Mary of Burgundy*, written by a scribe named Nicolas Spierinc and illustrated by the "Master." It had somehow made its way from Flanders in the 1470s to Vienna and eventually to Paris, as did the early fifteenth century *Très Riches Heures du Duc de Berry,* by the Limbourg brothers, certainly precursors of Van Eyck [1].

All present, with the exception of Petrus, recited a Latin prayer for peace that accorded with the spirt of the devotional gem, with words to the effect, "Da, Domine, propitius pacem in diebus nostis, ut, ope misericordiae tuae adiuti, et a peccato simus semper liberi et ab omni perturbatione securi…" (Grant unto us, O Lord, graciously grant peace in our days, in order to that, through the assistance of Thy Mercy, we may be always free from sin and safe from all disturbance.)

Their solemnity paused out, sending a chill throughout the great Church. Petrus felt the breeze. Noticed some tears. Bowed his head. His heart was racing. Then, Bruno suggested, "Perhaps our guest is weary from his journeys." And several people proceeded to provide him towels, show him the way to the restroom, and serve up platters of savory dollops and a fine Delft jug filled with pure cold water.

Then they left him to his own. All alone in the enormous church. He was feeling the effects of so much wine tasting. But more so, a sensation of human generosity that was, in its cumulative resonance, new to him.

"Do I need a key or anything?"

"Hardly," Father Bruno chuckled. "You'll enjoy the light in the morning, of this I am certain."

"I have a question," Petrus asked.

"Yes?"

"You have no electricity?"

"You *have* been away."

"Not for over thirty years, maybe longer, but who can remember such details," a woman among them—she had cooked the Ligurian garlic and basil rich pistou—in her eighties, answered.

Father Bruno added, "There was a satellite, or so it was rumored. The peoples' satellite. Open access. Confusion reigned, of course, as you must know, you are old enough to remember. But the social frustration"-.

"Chaos absolu!" chimed in another, revealing the deep-set anger.

"Yes, but it all eventually gave way to silence. There was no point in trying to fight it. One must know when to step out of the rabble. Survival instincts dictate disappearing, not in a throng, but in one's basement, or attic. And then everyone had to start over. In a different way. It was not so bad, not in Paris. I don't know about elsewhere."

"The main thing was to stay out of the direct Sun in the middle of the day," the blindman said, pointing upwards.

"And avoid stray bullets," declared another. "And hope that those stragglers outside the city would still manage to bring fresh vegetables into town."

"And now God has forgiven all things. The Sun is not so bad," Bruno concluded. "Anyway, one could not stay out of it forever. We are social animals, alas. We must try to help when we can."

He went on to describe a friend of his, a French physicist who had explained that certain sub-atomic particles were penetrating indoors.

"There was no escaping it," the blindman, Salvatore, added.

So Bruno began, "Much talk of building up immunity. Scientists worked round-the-clock to figure out a solution. All the normal trial phases for drugs were eliminated, of course. Nations, corporations, people – everyone was hurrying to beat the odds. Many simply never went outside. But you have to get your bread? Anyway, indoors, outdoors, those nasty little particles. It was no use."

"But what about the vibrant-looking animals I saw? From all over the world? Are they simply immune?"

"Yes, and no." (*Everyone seems to know about all this*, Petrus was rapidly concluding. Physics, particles, medicine, zoology. *Bizzarre*!). "The Paris Zoo was liberated and it is my understanding that zoos everywhere were so, as well. 4-D printing helped many with a plethora of injuries -rabies, Bumblefoot, sour crop and Frounce - to survive in newly fashioned silicon jell booties for raptors, double-padded loafers for bears, other ingenious prosthetics for a host of freed species, and vastly improved internal organs. I think they got quite carried away."

"Okay. How do you know all that stuff? You're a Priest."

"I first studied to be a veterinarian. Then I got completely sidetracked by the human condition," he replied plainly. "But while our human crisis unloaded many ancient biases, it also prompted a veritable Renaissance of religious beliefs, all culminating, both in theology and science, in degrees of compassion. That, and the experimentation by geneticists and others targeting animals and plants, but ultimately intended to change human nature for the better, gave me courage to carry on."

"You know about all that?"

"We all know. Where have you been?"

"Do you think there are still animal migrations coming and going? All around the world? I ask because, as a scientist, albeit a young one, there are several disconnects in both worlds of spirituality and all that stuff we do to one another, *for* one another, here on earth."

"So. You are a scientist. What sort?" Bruno asked.

"Evolutionary biologist, you might say. But I'm not unfamiliar with the Fermi Paradox. The Drake Equation."

"You'll have to enlighten us. I don't know them," Bruno said.

"Well, if you reverse both of them to focus strictly on life on this planet, not extrapolations about life elsewhere in the Cosmos, then there is no question of a survival factor. Yes, apocalyptic events happen. But that is not the same as some individual l'appel du vide or call of the void, but quite the opposite: la passion de la vie."

"I always believed there was life elsewhere in the Universe. There has to be. But I agree even more wholeheartedly that there is already life enough on this planet to sweep us off our feet for millennia."

Petrus was doubly perplexed. "But, not to get too technical about it, there must be a biological explosion, a great birthing like we saw in the Cambrian multiplier effect 540 million years ago [2]. How else to explain this multitude of species running all over Paris, playing, from what I've seen?"

"I don't profess to know. I'll leave the past to the great libraries, and curiosity seekers like yourself. All I can tell you is, all the zoo animals were liberated, right out of Genesis. The most recent past, in retrospect was the great evil. Our predecessors did all this. Now it is all okay. They must have done something right amid the many madnesses to engender a new generation of humans who embraced virtue. Curtains on the dark were lifted. Light shined in. Truly. A kind of welcoming laziness. The gorillas of Paris are thriving. Go and figure. The Bois de Boulogne must be the new Congo. No one wants to ever risk repeating all those mistakes of the past. And so, there you have it. Maybe we finally all learned from history. No one would dare harm another creature. No more fishing. No more foie gras mousse. No one will cast the first stone, never again I would hope, or tinker with a coil of copper wire for adverse purposes."

That set Petrus to wondering: "And I suppose no magnets?"

Father Bruno understood the boy's thinking. But he felt it better to let queries and deliberations fade out. He sighed, then declared: "None that appear to work, for whatever purpose."

Bruno was no fool. He grasped Petrus' line of inquiry and knew that this was a stranger to be cautious about. A dog whose tail could not decide which way to trail, for good or possibly ill. Nothing was as it seemed. He had to be careful, but the mendicant in him also sensed a true quality of irreproachability. Like a wild child.

"I remember a time of desperate how-to's," Father Bruno mused. "People tried, in the beginning. But your average Parisian didn't have a clue how anything worked. And honestly there wasn't the motivation. We were already two centuries behind, in that sense, finding ourselves in this incredible city that no longer had a single power source. No lights, refrigeration, communication." And he pointed upwards. "But thousand-year-old stained glass leaving its magical lights. Or to wander throughout the Louvre, without running into a single other person. The ability to bake bread, grow vegetables, leisurely stroll through the Marais, the Place des Vosges, one uninhabited mansion after another, with not a single care in the world. I don't know about other cities, countries, what they may be doing. But I suspect it is the same everywhere."

"If I may, what about all the corpses?"

There was a deep chasm of silence.

Then, "funeral pyres," Bruno plainly explained; "to try and contain the spread of diseases. Particularly Cholera, Crimean Haemorrhagic fever, mutant MERS coming from Russia, and new poxes, presumed to have originated on Malta. But absolutely no reason to attempt the impossible, that is to say, rebuilding civilization which, to be honest, did not work. We've got the best of it. The 'Mona Lisa' remains safe in her nook. No one will ever again steal her. And most of the horae, those lovely, illustrated prayer books, have been well looked after."

"The "Mona Lisa" was stolen?"

"But that was ages ago. You didn't know? By an Italian patriot who kept it near his oven in a modest little apartment near the Louvre for two years [3]. Can you imagine? Ancient history, long before all our time. Anyway, she's enjoyed a little makeup since then. She looks good," he smiled. "And the ancient trades, like baking, or planting corn, visiting, indeed venerating libraries, all that remains true to the values that first instilled the arts of reading and writing. Today there are readers, not highwaymen."

"No riots? Pandemonium?"

"Yes, in the beginning. Mobs in the banks."

"It was dreadful," winced Aminé.

"Just to hear it was bad enough," added Salvatore.

"No more ATM machines. Or any financial structures. The rich bankers fled. They had no idea that it was just as bad everywhere, or so flew the rumors. Class distinctions, finished. A new Sun King to blame. And blame they did. Then governments collapsed. Then it really got bad. All the televisions and radios, computers, hand devices, telephones, trains, taxis and airplanes stopped working for lack of any energy source, presumably."

"Radios as well?" Petrus pressed.

"Yes. And after everything went silent, and dark, and candles and matches were worth a person's soul, people set about to get on with it. They kept burning their dead. Riding their bicycles as long as the tires weren't flat. To become familiar with the old world. People are people. They can make do. Especially when all Paris is your oyster. Of course, we used to think that people became good after a disaster knowing it would be over as quickly as a tornado. This was not the case. It continues, as we all know."

Finally, exhausted, Petrus declared, "I need to get to Belgium."

"Why not. That should be possible," Father Bruno stated. "Though there is so much to enjoy here in Paris. Spend some time."

"What's in Belgium?" asked one of the congregated, an old man with many scars on his temple.

"My family, I think," Petrus volunteered. "And where I'm headed is not so far away. I hope to come back someday."

"And you shall always be welcomed, young man," Father Bruno said definitively, patting Petrus on the shoulder. "Amazing, simply amazing," he declared, gazing at the youth that was once himself.

References

1. See *Farnese Book Of Hours*, Ms M.69 of the Pierpont Morgan Library New York, Commentary by William M. Voelkle and Ivan Golub, Akademische Druck-u. Verlagsanstalt, Graz, Austria, 2003. See also, *The Hours of Mary of Burgundy*, Manuscripts In Miniature, Codex 1857, Vienna, Österreichische Nationalbibliothek, Commentary by Eric Inglis, Harvey Miller Publishers, London, UK, 1995. See also, *The Gualenghi-d'Este Hours – Art And Devotion In Renaissance Ferrara*, by Kurt Barstow, The J. Paul Getty Museum, Los Angeles, CA 2000. See also, *Le trésor des Heures* by Fanny Faÿ-Sallois, Desclee De Brouwer, Paris, France, 2002. In addition, *Les Belles Heures Du Duc De Berry*, Introduction by James J. Rorimer, Notes by Margaret B. Freeman, Thames And Hudson, London UK and Metropolitan Museum of New York, 1958. And finally, *The Master Of Mary Of Burgundy*, by Otto Pächt, Faber And Faber Limited, London UK, 1948.
2. See Erwin, D. H.; Laflamme, M.; Tweedt, S. M.; Sperling, E. A.; Pisani, D.; Peterson, K. J. (2011). "The Cambrian conundrum: early divergence and later ecological success in the early history of animals". *Science*. **334** (6059): 1091–1097. Bibcode:2011Sci...334.1091E. doi:https://doi.org/10.1126/science.1206375. PMID 22116879. S2CID 7737847. See also, Droser, Mary L; Finnegan, Seth (2003). "The Ordovician Radiation: A Follow-up to the Cambrian Explosion?". *Integrative and Comparative Biology*. **43** (1): 178–184. doi:https://doi.org/10.1093/icb/43.1.178. PMID 21680422.
3. See *Leonardo And The Mona Lisa Story*, by Donald Sassoon, An Overlook Duckworth/Madison Press Book, New York and Woodstock, NY, 2006, Chapter 4, pp. 212-259.

Chapter 23
Affable Landscapes

He had never seen a meadow, flush with sunflowers in bloom. Polders of Entsiedlung, the best soils he'd ever stood upon, nothing like in Antarctica, to be sure. And what used to be called sheep runs. Bastides and Seigneurial moats and windmills. Or geometrical landholdings dating back to Roman times. He was walking alongside a freeway. A cheerful woman, dressed plainly, a middling skirt and wide bonnet rode by with food in her woven basket. She stopped to enquire of Petrus, offering him some fresh kumquats which he accepted gratefully.

"Where are you going?" she asked matter-of-factly.

"Belgium," he replied, wondering why the French were so friendly, compared with the mostly fearful Spaniards.

Later on, in the welcoming cool drizzle, two men with a four-wheeled cart filled with sod, holding sickles, walked past him, saying nothing, no apparent curiosity at all about this tall, Scandinavian looking fellow of 31 heading northwest, wearing a rucksack crammed with a most eclectic assortment of emergency supplies.

Through the forests of oak and ash were scattered red-brick manors, a white limestone chateau with a vibrant looking vineyard, still providing; a huge garbage dump swarmed by gulls; and a deserted hamlet, wild beehives, and then what appeared to be an excavation. Barns empty, Red West Flemish bulls at their leisure, an ecclesia lignea, oblique supports, built in the traditional Stabbau manner, vertical joined timbers, with a little boy playing some kind of solitary spud—a little boy, the first!—near the unmaintained entrance. This kid was no older than eight or so, Petrus figured. He approached the lad whose dark skin, perhaps he was North African, apparently abetted his survival in the burning Sun. To see a boy was shocking, by this time.

"Que est votre nom?"

"Diderot."

"Tout va bien, Diderot?"

"Oui, merci. Que lest ton nom?"

© The Author(s), under exclusive license to Springer Nature Switzerland AG 2022
M. C. Tobias, *The Maiden Voyage of Petrus van Stijn*,
https://doi.org/10.1007/978-3-030-97683-5_23

Simple contact with the child. He seemed altogether normal, an orphan, perhaps, of the new world, or maybe not. But he had no demonstrative need that Petrus could decipher; not a stray dog wanting to follow him. Diderot had his own business to attend to. Wandering off.

Amid a vast sluicing acoustic in the air, of millions of crickets communicating. And then the strangest of all chance encounters: a real giant Moa, browsing on the groundcover right beside the D932A. The freeway was half-polka dotted in huge dandelions, buffalo sward, goat willow, and blackthorn.

As he paced himself, he continued to feel now and then a deep, fulminant pulse, something other than crickets; between a temblor and the sense of one's heart stopping for half-a-second, the rush of blood audible to one's ears. Pushing slightly against the chest. An indescribable sensation that permeated the world with a half-imagined thud. For months he had forgotten about it. A heart arrhythmia? He'd never been told about it, if that's what it was.

Creamy white and mottled brown wild horses grazing at leisure in the forests of ash, larch and aged twisting yews. As he walked, mile after mile, greenways had spilled over. Amid high thickets off the freeway, he came upon a gas station retaken by ebullient vegetation. He examined the emptied shelves and long-broken down exterior. It had become, sometime in past decades, a forsaken fossil, smelling of dead rodents. Vagaries of weather had rendered the entrance more Middle Ages motte-and-bailey. There were gulches and rises, no easy way to drive in over ruptured earth re-invented by the taproots eager to communicate with one another between triangular piles, shamelessly plundered. The entire building had been outcompeted by the irresistible force of surrounding heaps of wild raspberry and thorny liana. Not the stuff of commentary or chroniclers. But of dimmed traditions and the normal succession of biotic communities eager, some more than others, to face the brilliant Sun.

The Sun, which shone between bottlenecks of storm, looming black lenticular clouds, oddly ashen. Multiple rainstorms that had drenched Petrus, propagating deep fog in their aftermath, before the solar shrapnel would again pierce the odd meteorological landscape. He had seen similar, though vastly more violent mood swings where he had come from.

Far above, miles away atop a densely forested ridge was a castle, white and from afar splendiferous. As he climbed up towards its ensemble of opulence, the most apparent fact was its lack of doors or windows. Up the curving balustrade upon Aubusson runners, with Rococo handrails rusted and in part removed, Petrus wandered from lavish room to room, flimsy chandeliers still hung upon chains. The embroidered silken walls were intact, though the shredded draperies were now home to large festoons of busy webbing cloth moths, aggregating all the way to the summit of the Mansart roofing, which was eaten through. Broken slate lay upon the centuries-old parquet flooring of oak. And broken glass everywhere. There were still largely unblemished paintings on the wall, even a Korean cell phone adjoining a Chinese computer in an office room with Gustavian imprimaturs on swivel chair and table, down the carpeted hallway.

Upon a lavishly manufactured Louis-the-something table, was a newspaper, dated October the eighth, 2035. Stained from leaks from the roof, opened to page 3, a paragraph that at once caught Petrus' eye: *Le ministère des Affaires extérieures avertit que les frontières avec la France ont été fermées et que la disponibilité des salles d'urgence dans chaque hôpital est dans un état déplorable. Il n'y a plus de traitements disponibles. Bien entendu, le gouvernement tient à exprimer sa profonde douleur face à cet état de choses...*

Closed borders. Hospitals overrun. The Government sends its condolences to one and all...

Petrus handled the telephone, the computer, just to see if there were, perhaps, anything like a signal. Not a chance. The entire blank stare of the chateau penetrated every last corner of the once lived in fortress. Occupants had sometime in the impossible past enjoyed a life here, to be sure. Was October, 2035 the date of annihilation? Petrus wondered. When one way of life was finally evacuated? Adj or pre-adj? It didn't matter.

On a sofa lay the parts of a vacuum cleaner. Someone had tried to fix, or rebuild the motor, commutator, stator windings, and electrical connections. Clearly, whoever it was had a dynamo in mind. There was dried blood on a Phillips screwdriver.

A levant was blowing in through the broken windows, carrying with it a turn of weather, dense with moisture. Petrus went back down the stairs, and out the open entranceway.

Tulips and climbing Bourbon Roses were blooming within the private cemetery, and the gravestones could be deciphered amid the flourishing mosses and lichen. Jeanne de Brunn, Sceaux de Mainfroit, Jamea Mainfroit, Madame Eques Charbonniers... Cherub sculptures joyously bound to the central cenotaph, upon which was ornately etched, *Est mort ici en paix*; and *Bonne chance au monde. 2004–2036*. And standing to the side, a dolorous angel, face withheld, weeping into her robes, sculptured in a wet grey stone already budding with lichen.

Four does gathered near, picking at the grassy banks of a troubled moat, where peacocks knelt down to drink from the gathered ponds, their otherwise glittering primary feathers soddened with mud, he noticed.

In the distance, for just an acoustical flash, he thought he'd heard the most welcoming distant braying squabble of penguin. But it was only lightning accompanying an auroral electrojet, a visible geomagnetic storm high above the cloud-soaked arc of France.

It was after five and the rain had started again. At first, Petrus thought to stay out and sleep under a dense grove of trees, covered in the abundant blankets—thicket, holly, spider web and mats of colorful leaves—of early Spring. He knew nothing about European ticks, or the risks of Lyme, or Babesiosis or Tularemia. But the rain now turned to hail and he thought better of it, in any case; returning, through one of the entranceways, and finding a lavish canopy bed in which to sleep upon the second floor, down a long hallway teeming with art. He noticed among the signatures a Jules Dupré, amid a pile of attic-like reminders from another age, much of it host to mice droppings. Bigger than mice, but not quite rat. Muskrat? Possum? He had no clue.

Before long he heard—he was certain—wolves or thoughts of wolves, transmitting information from within the dense surrounding forests of the Compiègne, and d'Halaate, as they were known to anyone who might be local. The calls were gathering force across the Forêt de Chailly and Petrus imagined, as the wind continued to pile on, breaking old dry bones of former structure and sense, across the vast summit of the chateau, that the wet stormy levant, having turned to a mistral-like gale, had brought indisputable joy. *One could live here*, he divined in the grace of the moment.

Chapter 24
Coming into Focus

"Two New Friends in France," Photo © By M.C. Tobias

Petrus had grown up a biologist like perhaps no other person. Without a single tree to touch or be tutored by for three decades, he knew enough from the single-minded pursuit of all that transpired at pes and throughout the thousands of square miles of Queen Maud Land, that a Pedunculate oak was subject to hybrid powdery mildews, epicormic shoots, and defoliating insects. But he also understood that there should not be lemurs moving about the Wych Elms and sequoia. A pollarding had taken out the high branches of cedar, cork, green alder, and lime-trees throughout the estate, while trenches like those used in wartime, or by archeologists had also been built and later abandoned. There were skeletons. One room in the chateau contained a corpse of a child seated slumped upon a dark velvet bench. The skeleton had not been picked over, but covered in a silken, embroidered blanket. Who was this person? How had he died? What were his final words? Did anyone comfort him in the end? Important questions. Unanswerable. But the universality of such comforted death was unequivocal.

He wandered the unbounded estate. Discovered two sloths sleeping in the crook of a mighty chestnut. One of them spotted Petrus, and fidgeted, but only slightly. Overhead, an unassimilable flock, or flocks, of Greylag geese, no doubt from Sweden, were noisily honking as they progressed in a hurry toward not southern but eastern reaches. It made little sense to Petrus.

It must have been late February. He had not kept careful note of passing rhythms. Everywhere, hybrid acoustics were crowding a newly revealed biosphere.

Petrus remembered from childhood some of his mother's colleagues from the archipelago comprising the Îles des Pétrels, at the Base Dumont d'Urville. That albatross specialist, Jacques had evidently died of an unqualifiable virus. A suspected zoonotic bird flu transmission. The French self-quarantined an entire Winter, then decided it was safe enough to commingle once again with colleagues throughout Antarctica. There had been multiple instances of such communal superspreader *choices*.

There was no use of a toilet in the Chateau. No water, or water pressure. Darkening yellow rings and mineral crystals within each of the dozen toilet rooms. Handles flopped with a metallic empty sound, producing no force. Food supplies were there, in the largest industrial kitchen, including sturdy, fungus-greying rounds of Camembert de Normandie, the same blend he'd sampled in Paris; a bloomy rind cheese. In one of the desk drawers, gnawed on it appeared by a beaver or rats, an unwrapped candy bar. In cupboards, stockpiles of last-ditch effort items, including tins of white asparagus and cured anchovies. Petrus searched out other cabinets in the kitchen for a can opener. There was only a power tool to accomplish the most rudimentary of initiatives, but its batteries were dead and encrusted.

He scoured the numerous outbuildings, a barn adrift in old tools. There were jumper cables, animal halters, a rare early eighteenth century Grand Figural Cartel Clock after Gilles-Marie Oppenord's work from some Bâtiments du Roi, a clock several hundred years old that was frozen in its time.

All the family pictures might have well been taken on holiday somewhere in the Morvan or Fontainebleau, or just out back, for that matter. So perfect were the landscapes. They resembled daguerreotypes, like those images by Joseph-Philibert

Girault de Prangey who would produce an after-effect of the first non-human animal (a Roman steer); and the legion of later photographers who had explored polar reaches, plastered on refectory walls for posterity at Nordenskiöld Base. Uneasy remembrances.

Without the least schedule, Petrus manufactured in his mind a date certain to leave the premises. The ambient air temperature was quite mellow. The hail had collected in inches all over the unmowed carpets of grass well-tended by a colorful company of Pyrenean ibex; a soft riot of ground cover, a species of vinca, and crawling subgenera of Micranthobatus, of the genus Rubus. Black and blueberries in abundance. Jars of mustard in the kitchen. Someone had employed a little industry of pressure canning, both jellies and high acid foods. The jars, all labeled with an un-addled handwriting, sat within preservative shelves, lined with linen that had not been touched in years. In all, there seemed enough food supply for Petrus to do nothing for weeks, as he reconnoitered his oddball future and the itinerary driving it.

He awoke early on the fourth morning having heard, daily, the faint comings and goings of a horse-drawn carriage and several bicyclists. Petrus knew nothing of such sparse traffic, or the normal rate with which bicycle tires lost air. Only that it was happening. There they were. As if the world were simply carrying on.

He felt strange amidst these odd machinations of people inhabiting a low, wet, and verdant space, where every horizon was obscured by dense glens of forested verge and blur. For one transitioning between two homes, two exiles, all these wild stands of maize, mustard, of blooming black hollyhocks, spoke of both data base and confusion. The chatter of invertebrates roaring across the patchwork of terrestrial habitat. The aerodynamics of biomass on its infinite appointments of tack and gybe, all thoroughly disorienting. Like a berserker Borneo of biophiliac messengers, choreographed by some miracle enunciating its imperial priorities, despite what was an entirely random fanfare. He was either losing his mind, or rediscovering it, this eerie orphanage of wedded ties that had abounded in the face of what appeared to be a hybridized evolutionary élan vital.

Beneath it all, an illimitable tomb that had swallowed almost the entire remains of humanity.

Quite honestly, Petrus did not know how, or why to relax amid this bounty of biological energy and distinct melancholy. What could he know? Nothing of it, but by way of his own strange past, now. They had somehow urged him into a cognitive frame of top pleasant memories. Rejuvenation.

He wanted this new life, all of it. His fingers, face, whole body into it. To become it. Hugging a tree's rude spikes and dusty bark, lying down in the thick mud of meadow and marsh.

The property had no boundaries. From its rooftop Petrus spied other chateaus far off; a neighboring hamlet where not a single bipedalist stirred. The weather had turned uncomfortably warm, humid, mosquito infested, and he thought how involved his mother would have been. Were these natives, invasives, hybrids? Overwinterers? Fire and scentless plant bugs. Carpocoris stink bugs. Odontotarsus family jewel bugs. Treehoppers and spider wasps. The intensity of these age-old networks made him half-queasy with their complexity. The very biology was just shy of inducing an unpleasant dizziness. He was not entirely comforted by the splendor of so much. The sea was still seething in his ears, timed to his racing pulse.

Every one of the individual insects and spiders was a multi-million-year-old native, or were they? Tiny chimeras, perhaps, invented by the likes of his own mother?

But there was no doubting now that Petrus was the intruder, a convoluted outsider. This apparently rude juxtaposition did not make for easy assimilation on his part. His senses were alarmed, his own disfigurement—so ungainly a giant—was not a little frightening. He possessed neither the know-how nor keys to making peace, as the earth's biodiversity had outpaced all previous borders and watersheds, taken on a multiplicity of personalities whose aboriginal and unfettered blood flows vied with his own cognition. Herein lay the fundamental collision of consciousness with a world that was perfect without more than a couple of human beings. He was biologically de trop. Petrus felt useless. There was no task at hand. Nothing to be done. No cause to pursue, other than the vague hope of a reconnection. It was weird.

Belgium began to feel, in his mind, like the legendary Ithaca, or El Dorado.

Eventually, by around noon, at the hottest time of day, as humidity visibly rose up from the ranks of crops long ago turned back to the wild, he repaired to the canopy bed and lay atop shaken out pillows of silk and down, just thinking: What should I do? His loins stirred with the memory of the Spanish beauty, Dulce. The heavy loss of so perfect of (a pointless recollection), amid the stowaway chaos of his voyage. The disappearance of the entire history of research data. All his friends, his mother, everyone he grew up with. The absence of everything. And no one with whom to speak of it. How to even convey the scope of what had vanished, that last spark, in the muddled event horizons of his tenebrous Antarctic past? How to accept that it was even a *past*? No longer the present.

He lay there in the surreal evacuated luxury of dead strangers, and realized that it was all on him, now: He had to decide. To make his next move. To pretend something was at stake. That life must go on, at least for himself, in one fashion or other. In a place once known as Belgium.

Chapter 25
Going Home

The decision was made the moment he started fearing about the odd passing stranger. With binoculars he could spot the occasional rifle standing beside the driver of one of the carriages. This was obviously a not uncommon transit area. And, in fact, it was the main highway between Paris and Brussels, the chateau sitting no more than half-mile from the frontage road. He took the largest knife he could find in the kitchen drawers and stashed it in an easily retrievable nook within his ruck-sack. Then bid adieu to pictures of the resident family, and ventured out into the wilds, his idea squarely fitted with an image of that land to which his soul belonged. He had inculcated these rural uplands over decades, from so many half-intimations and loosely applied reminiscences that were not *his* actual memories.

A Bruges, or Brugge, that existed in the vast, pleasant avuncular time, linear and simple, festive images of gift-giving, of Xmas and Summery strolls through a maze of brick-lined canals. That was the residual vortex where black geese glided effort-lessly, in charge, before everything in Petrus' life had happened. That was the ances-tral land he had assembled from so many anecdotes and depictions.

There was no border, nothing to suggest he had at length entered Belgium. No markers from the former history of the European Union to indicate a geographical change. Nor did Petrus have any idee fixe that might situate any palpable coordi-nates or namesake; no church, no village, nothing to say, this is Belgium, this is where your family is from.

It was the scent of salt air that reminded him of something he could not possibly remember. The coastline at Knokke-Heist where trekpaard draft horses were ridden by children and helium balloons disappeared out over the waves. It was a narrow barren of a beach, dozens of miles long, punctuated by sinking sea communities that had nothing historic to declare. Knokke itself, Heist, Albertstrand, Zoute. Places with cultural centers and seafront hotels, abandoned, half-or-completely underwa-ter. A kite broken and jutting from upland shrubbery. Scraps of debris here and there further inland from the flooded estuaries. A Red Bull can, an old anchor, a deterio-rated, drenched dock, dead fish in gaseous eddies lingering about piles. Oil pains

M. C. Tobias, *The Maiden Voyage of Petrus van Stijn*,
https://doi.org/10.1007/978-3-030-97683-5_25

streaking the gullied sands. Shards of razor-thin glass. A dismal admixture of insults where perhaps a century or so before, there had been a thriving intersection of humanity's age-old communions with the sea.

He headed inland, instinctively guided by the bicycle trails along the canals all leading towards Bruges. He knew it to be so. The scent of no trail in particular, but of the direction. Overhead, storm clouds were brewing quickly. The temperature kept fluctuating. Muggy, then chilly. Crazily unguessable. Never in his life, other than in the sauna at pes, had he ever enjoyed such mellow ambience in the air. A tropic-like of cancer 1800 miles too far north. The cancer itself, which was said to have explained everything, Father Bruno had intimated. Blind children, blind parents, blind horses, blinded birds, burnt amphibians.

And just as many or more who had somehow built up an immunity.

But as he walked down the N31, or what once was, through Lissewege and Kruisabele, he seemed to know precisely how to find the Belfry, passing through the Park Winkelcentrum, Zeveneke, then reaching the Markt. He knew it all by some heart that had been planted prior to his life. Now re-transplanted in the craving center of a new identity. Every chance encounter would matter. Seeing hippos bathing in a canal, egrets and swans on their backs, making no sense, but every sense. After all, who would have imagined that the tyrannosaurus rex once hunted in packs throughout northern Europe. Or that all the shops were once open, and tourists arrived in gas guzzling buses, children with nothing to fear. Neither chocolate nor French fries would ever dry up.

What was clear from the corner of the Gruuthusestraat leading into the Dijver, over the canal, and crossing the Groeninge at the old Modeatelier La Rose was that, somehow without substance, he was standing in the arena of his homeland, his very genetic comfort zone. But did not know, feel, have the least inkling, what to do. No one to speak with about anything. Not one shared experience. He stood there alone, a little out of breath.

It was a magnificent place.

Chapter 26
A New Life

This was a surface scene with no technology. No lights, cars, only half-submerged tramway tracks, vaguely the rare and withered overhead wires. No sign of power. People, there were a few. Wandering about their business, some looking at Petrus and wanting to say something but not daring to do so. Did he look so different? A bicycle here and there.

When Petrus' parents had conceived him the town's population was under 120,000. Presently, he was soon to discover, Bruges numbered slightly fewer than 35 residents. No one could be sure of a census, per se, because a few never came outdoors and no one, or small group, had ever taken it upon themselves to break into every structure searching for bodies, living or dead. 30–35 intensely fraying stories to convey. Not unlike penguins, these upright denizens of the late twenty-first century had been stressed beyond all concept of comfortable evolutionary latitudes. The biologist in Petrus thought about genotypic expression, the nucleic acid polymers, and organic molecules whose eukaryotic legacies had been so rudely manipulated, all within a matter of decades. Could it have affected people?

Cell signaling, the four nitrogen bases of DNA, phosphate groups, and 5-carbon sugars whose combined amino acid chains had become the rage of a human arrogance that no longer cared about the perils of re-evolution: these were not easily detected transformations. His mother was certain it was all for the best. Seeking instead to catalyze real-time change, not waste a second. Bypass the million-year rule to accomplish in vivo, or in vitro near instantaneous causality. Crude, bold, disastrous. Not even a decent UV-protective skin cream resulted from the frenzy. What Petrus had seen thus far were the two profound stanchions of change: a depauperate human population, and a thriving, if fairytale-like conflation of biodiversity alterations and abundance.

And what else? Petrus sat upon a park bench. From there, and presently, he now took stock of a very down-to-earth reality. He sat in the same waterproof refridgiwear he'd had on for many months; thermal socks, tennis shoes given to him in

© The Author(s), under exclusive license to Springer Nature Switzerland AG 2022
M. C. Tobias, *The Maiden Voyage of Petrus van Stijn*,
https://doi.org/10.1007/978-3-030-97683-5_26

Gibraltar in exchange for his Antarctic waterproof boots, a neck gaiter, and beanie. Even a green hooded jumper.

By noon the temperature was in the 70s. For thousands of miles along the way he had pondered this illogical climate. Nothing he had ever studied claimed to understand what was truly happening, except that atmospheric, and stratospheric rivers of change were to be expected. There were any number of scenarios to explain how the different geological and biochemical components of a complex system were reacting; entropy and negentropy vying for some attained stability that was not to be.

Petrus felt feverish with heat and became quite anxious. An ancient yew tree creaked above him, where he sat beside a row of large-leaved lime trees spreading along a narrow pedestrian path. Birds were chatting up a European beech, flying back and forth between Sweet chestnuts and Persian walnut trees. Petrus had never seen any such species, of course; neither the avifauna nor the plants. One Pagoda tree in particular, *Styphnolobium japonicum*, he thought he recognized from a biology textbook. The park contained Silver maples and Turkish hazel; Black poplar, Bigleaf maple, Japanese larch, and a variety of Quercus oaks. Trees from everywhere. And birds as well. Cuckoos and wrynecks. Green and Great Spotted woodpeckers. Kingfishers and collared doves. Even the exotic Ruff could be seen beside Eurasian coots and common Moorhen. Ruddy Turnstones, flying away from the sea, had come to his feet, along with gulls and pigeons, in hopes of scraps. And how difficult was it to give a scrap of bread to a pleading bird? Very, if you did not have bread. Petrus was hungry, as well.

Small copper butterflies commingled with colorful fritillaries. A pale tussock moth made its plodding way along the limb of an ash tree; grey wagtails were prolific, as were noisy dunnocks, carrion crow, the Eurasian Linnet. Red squirrels were everywhere frolicking in the park—*was it their mating season?* —but their lack of territorialism sounded a clarion signal that demarcations had long been abandoned. An enormous brown hare wandered into someone's vegetable garden, where hedgehogs had already been at work. In the canal waters, Petrus saw movement, then noticed the Common Bream. An insect landed on his finger, the long-bodied *Clistopyga incitator*. A beauteous *Dromius quadrimaculatus* crawled by. At least he knew some of the residents already.

There were slow feeding dolphins in the pellucid canal waters. And they were happily communing with porpoises.

An elderly woman seated herself near to Petrus, to feed the birds. She proffered barm cake and kept looking at Petrus. Her face was all but frozen in time, as if some sudden paralysis had made her rigid, freaky looking.

"You?" and she shook her head, as if something totally incoherent and without precedent had shaken her shoulders.

"You are who?" she let out, with a gasp. And then offered Petrus ample slabs of baguette to hand out. She carried the breadbuns in a bag expressly for the purpose of those clustering about her feet, flying onto her shoulders, and atop her head.

"Yeah. Bedankt," he replied instinctively, taking a handful and greeting hungry pigeons now struggling to stand on his shoulders, as well, and in the long dusty hair on his head.

Within minutes a male friend of hers arrived and sat beside the bird feeding woman. He too shared something of that same paralyzed stupor.

"Ghoeie middagh Ghoeiendagh," he said.

"Hey. Hoed, e me joen?" someone within Petrus replied effortlessly. But he did not speak the language. Could not really utter a single word without giving in to some reflex, a time when he must have heard phrases from infancy. Onsa had mostly spoken English and French. "Klappe ghy Ingels?" Petrus went on.

"You look so familiar to me?" said the elderly man. "Have we met?"

Petrus shrugged. "I am newly arrived, just today."

"New from where?" he persisted.

"Antarctica," he stated.

There was a commotion of stilled incredulity. A swarm of passenger pigeon-like memories.

"Come on now?" the man, probably in his early 80s questioned. His beard had been populated by reddish streaks that all but glowed under the intermittent glare of sunlight.

"Yes. It's true. I've just come from the far South."

"So far South?" a lock opening in his mind, a door to the most remarkable anticipation.

"I was based at the Princess Elisabeth Station."

Now a fantastic set of prospects enveloped the elderly man, the greatest scientist left in all the remaining Flemish community. He hardly knew which sentence to put in which order. Then emerged a linguistic sensation, learned from his brother. It came from a famed manuscript in particular, a page of which began with the letter D.

"Domie labia mea aperies, Lord, open my lips," read the manuscript before Chancellor Rolin from some seven hundred years before, as painted by Jan van Eyck. Such were the words the old man reiterated.

"There is news from there?" Hans -his name was Hans - asked in an almost terrified voice.

"I can tell you everything," Petrus said, acknowledging a mutual anticipatory anxiety, speaking English now.

The man thought deeply for a moment, a devastating, hopeful conviction rising in his thoughts, one he had not dared entertain for decades, it appeared to Petrus. "Did you ever hear of an Osna van Stijn?" he wondered aloud.

"She was my mother." Petrus began to tear up.

Chapter 27
Rights of Passage

There was a most unaccustomed commotion, emotionally sprawling across the canvas of a park at midday in the middle of a mind-stream, a nostalgic punto fijo, as a Cervantes might have thought of it. Here was the original et in arcadia ego, a dazzling garden of Armida where the dumb learned to speak, knights and other heroes came to express their devotion, and maps were reassembled. A city park sprouting in every quadrant all the livre des merveilles du mond; affixed within a single Perdita, a lost, imaginary paradise, somewhere abutting a European Hyperborean center. It was, for Petrus at this moment, the true summation of all those tales of Coquaigne, where a Theuerdank met his God who might grant him peace in *this* world.

I always have, an inner voice reminded him. *I have always lived here...* A sonance of true comfort that now had morphed into that of the elderly gentleman on the bench commending this chance communion.

Is it possible? the old man mused, then: "What is your name?"

"Petrus van Stijn."

"van Stijn?" the bird feeing woman repeated. "Of old Stefanus?"

"My father," Petrus said. And hesitated to ask but had to give in to it: "Is he—"

The woman looked at the old man who gazed most sympathetically upon Petrus. "This is Mrs Sophie Jamar. Quite the politician, once. When there were politics."

"Nice to meet you," she said, neither she nor the old man answering Petrus' query. She gave him more bird bread. "It's got olives."

After a moment, "Petrus, dear boy. I am your Uncle, Hans. Your father has been dead for many years, buried just behind his house, or castle more precisely."

Petrus brushed the moisture aside from his left eye. "Hans van Stijn?"

"Yes. Your father was three years older than me."

"My mother spoke of you. You are an astronomer or something?"

"Actually, I *was* a cosmologist."

"Give him a few beers and he is still a cosmologist," Sophie smiled.

Finally, Petrus asked, formally and as if for the record: "What happened?"

"Since when?" Hans muttered. "You're asking, what happened to the entire world? In well over half-a-century?" But Hans van Stijn was still incredulous. Just gazing upon the lovely blond exile from outer space. "But how? How did you get here? We thought it impossible to ever lay eyes on you or your mother again?" Hans was deeply affected, just on the point of breaking down.

© The Author(s), under exclusive license to Springer Nature Switzerland AG 2022
M. C. Tobias, *The Maiden Voyage of Petrus van Stijn*,
https://doi.org/10.1007/978-3-030-97683-5_27

All three were quiet for some time, Hans shedding tears, Sophie as well.

Finally, Petrus launched at some length into his tale, mesmerizing his two listeners, interrupted only on occasion by the odd passerby who of course was deeply acquainted with both Hans and Sophie.

A wind, with it pouring rain drove the three of them into a firelit café nearby.

"No electric lights?" Petrus asked from wanton habit by now, knowing the answer.

His Uncle thought, you *have* been away, and shook his head perplexedly.

There were at least half-dozen or so persons who had come in out of the downpour. Drinks all around. Lanterns with real candles providing the light. Only vaguely effective coolers for the beers. Chocolate and loaves of bread in the cabinets, along with old rounds of cheese, preserves, and condiments.

"A stranger!" an elderly fisherman-type said with a craggy voice, peering unselfconsciously directly into Petrus' shyly averting visage.

"My nephew," Hans stated proudly.

"From *where*?" the fisherman went on, his roughness turned to astonishment. "And so young?"

"From here, originally," Hans said with a touch of excessive authority, to put an end to the matter.

"Lars is into everybody's business. One just ignores him," Hans said after Lars had gone on to accost a few other bar mates. "Anyway, he lost his entire family. Everyone you see here more or less did."

"Better to be kind," enlisted Sophie, licking the froth off her glass.

"Lost how?" Petrus asked.

"Young man, how old are you?"

"I'll be thirty-two whenever it is August."

"Thirty-one, astonishing," said Hans. "Adj or pre-adj?"

That took Petrus by surprise. He hadn't thought of it in a long time. PA, he said. "Of course, adj, pre-adj, does it matter?"

The storm was now pelting the café, slamming the windows, gale force.

"This is common in the Spring," Hans acknowledged, circling round his nephew's query.

"Spring? What month is it?" Petrus inquired.

"Spring is Fall, Fall is Summer, or something like that, at least that's the answer a physicist will give you. Adj. So you do you know what I'm referring to?"

"As I said, yes. In adj I would be something like 58. But I've really lost the ability to distinguish such things."

"That's probably about right. But it is all different, quite different than when I was young," Hans said aloud, but mostly as a muse might reminisce in an endless equation of sighs and alas's.

"For how long, such a dramatic shift in weather?" Petrus pressed.

"One never knows. Usually, it's quick. Bird mortalities are high, though. Poor little birds."

"And floods?""Oh yes," Sophie chimed in. "On occasion the Rozenhoedkaai has risen twelve feet above the walls of the canal, just last year. But the second and third floors of houses remain entirely habitable."

"There's mold. Much of West Flanders is now semi-permanently flooded," Hans added. "But we are told it is much worse in Amsterdam and St Petersburg. And Venice."

"What kind of mold?" Petrus asked. He knew Aspergillus, Acremonium, Fusarium. They had found their way to Antarctica, and he noticed plenty of them in both Spain and France.

"What Venice!" Sophie sighed. "That's the question. The Italians with their big expensive MOSE barrier project could never fix it."

Hans just shrugged his shoulders. He was no mold expert. Nor had he heard anything about Venice in many years.

During a pause in the smashing rain and hail, helping himself to a second beer, they were all poured generously around, Petrus asked, "How did my father die?"

"In his bed. He took the pill, of course. The cancer had advanced. He might have gone on another year. But he was tired of it. It's exhausting, obviously, surely you know."

"What pill?"

Sophie glanced at Hans.

"Son, the euthanasia pill."

"Phenobarbital?"

"Oh, much more predictable. A chemical algorithm had been deployed years before that, in one capsule the size of an ant, induced hypoxia, and rapid pleasant good-byes. First Switzerland, legally, that is. But then there was market contagion as global conditions worsened. Guns can be messy, which isn't to ignore the fact many prefer them. Part of it is just the anger. Frustration."

"Like a kind of revenge, but upon oneself," Sophie acknowledged. "Politics at its greatest efficacy. I know that sounds a bit harsh."

"How old was he?"

"Eighty-seven plus some months, but that's pre-adj. He was always the better looking one. But not in the end. He'd had a stroke, you know, in addition to other fatal conditions."

"I didn't know anything," Petrus stated.

"How could you. Of course, you didn't. I'm sorry."

"Yah. That's right. There was no way to reach me when my own husband died," Sophie added. "He was in Paris, I was in Brussels. I rode to Paris, eleven days on a very skittish horse, and another three to get to the truth of it."

"We called it the Carrington 6," Hans added. Then he realized that Petrus did now know, though surely would have felt it. "Carrington 6, Sol 2091-8-4. It was the largest coronal mass ejection to date, one-hundred-thousand times that of 1859."

"I don't understand, exactly?"

"Surely you saw it from Antarctica? It was the last vast flare that destroyed any hope of rectifying the burnt-out electrical grids. I'm not speaking of mere outages. The geomagnetic storms persisted for many years, off and on."

"Yes, we, the small group of our survivors at pes, Princess Elisabeth Station, we knew aspects of what was happening, certainly."

"Then you know that the storms no longer obeyed solar cycles. None of us ever understood it. When the Universe changes rules, throws mathematical constants out the window, it is disconcerting, to say the least. But the Sun may be turning into a small mass white dwarf star, who knows. And long before any of us could have predicted. For the non-physicist, as well as the physicist, that means, well, possibly end of life. But that's not what's happened, not entirely. In fact, quite the opposite. So, we all repeat: who knows!"

"Once he gets started," Sophie added. I'm just warning you." She grinned, like elliptical philosophical cigar smoke, dissipating upwards.

"Got it," Petrus grinned.

Hans was not deterred. "95% of all cancers occurred during that period. Estimates ceased to have much meaning, but we know that many plants and animals died, or migrated too rapidly. No food for them when they'd arrive, wherever they thought they were going. But then, with the magnetic poles so disjointed, let's say a locust heading for Algeria might have found itself in the remains of Kiev. Those that did manage were the ones with the highest gene pools. Like most mushrooms, they went crazy. Mushroom quiches of every variety. But one has to be very very careful. Obviously. Some things don't change. I've always feared mushrooms. And I must add, the genetics of everything following huge manipulations, both in vitro but also in vivo, were dramatic. Instability, lack of cohesion. These were our new compass readings."

"That's what my mother was in to."

"Yes, your father told me. Protein engineering, right?"

"Yes."

"Well, she was one of those pioneers then. It all worked out, in a way. Entirely new, synthetic biomolecules, site specific metagenesis, that's one of the descriptions. There was nothing that couldn't be manufactured in a test tube. No genome was spared. There were no rules, as it turned out. And if one country tried to institute some ethical norm, there were ten other countries eager to out-compete such standards, for profit and, in my opinion, the sheer lure of unfettered hubris. I'm not pointing a finger, mind you. Our species has always been power crazed."

"Of course you are pointing the finger," Sophie carped like a sour old schoolmarm. "Which is okay. My granddaughter was not careful. I point the same finger. Lots of them," she admitted.

"Careful of what?" Petrus pressed on.

"UV, principally. The coronal mass ejections," Hans continued. "They changed the earth. As you no doubt have seen, there are few people. The animals are all mixed up. The earth's poles not only switched, but continued to zig zag. Everything about the atmosphere was thrown topsy-turvy. Naturally I blame the politicians." He smiled and gave Sophie a little pat on the shoulder. "Not you, all the others."

"Come on, Hans. We listened to the scientists, but the people weren't entirely wrong."

"The people, which people?" Hans said heatedly. "Nobody listened to anybody. And those who did, what were they to do? Words, actions, none of it mattered anymore, other than the attempt to hold on to decency."

"He's absolutely right," Sophie conceded. "The situation was simply untenable. We have never been challenged, not in the social realm, every country, not like this. No institution to deal with it. Children all saw it coming, though. They just couldn't accurately describe it. What they felt. What we all felt. There was plenty of blame to go around. Until blaming became too tiresome. Everybody was exhausted, scared. Those who had faith feigned stoicism, but in the end, most succumbed to the torture of the golden age of ambiguity. This is what it was going to be. It just took a generation or so to come to really appreciate it. Worsening conditions perpetuate worsening conditions."

Hans continued to elucidate, intent upon conveying a kind of wrap-up that should convey to this astonishing newcomer the sheer magnitude of everything that had gone on.

"Without electricity," he pressed, "you can't imagine how difficult it was for all but a minute number of the very rich. And even they fumbled about like idiot children. No one could rebuild anything. All that information had been so sequestered, printed in the manuals nobody ever bothered to read, that when they did, it was at once apparent that you could not build a car, or a lightbulb, let alone a computer, from scratch. A hula hoop, boomerang or lasso one can fashion. Even a bridge or charcoal. But a wheel is very difficult to make, let alone an electrode. And with no technology, no knowledge skills, most were left with only the complaisance that comes from decades of engineering entitlement. Okay, there were exceptions. A lens grinder. The handyman, local mechanic, even the odd bicycle repairman with training in pumps and pressure, and naturally a pharmacist."

"What is the human population?" Petrus interrupted. "I've gotten different answers from different people."

"We think no more than 25,000."

"Truly? Worldwide?"

"Yes. And perhaps fifty individuals or so on the moon. Another two dozen in the international space station. But they are totally irrelevant, if they're even alive. The whole picture is pathetic, amazing after so many thousands of years. But there is this silver lining."

Chapter 28
The State of the World, Cont.

Hans continued explaining to Petrus what he could of the current conditions on earth, helped along by some stiff drinks all around.

"We have it on authority that distribution of basal and squamous skin cancers were nearly 100 percent in the equatorial zones of the planet. And every form of melanoma in other latitudes."

"Yes, many in the Antarctic bases died of cancer," Petrus stated.

"Most mammals were seen to perish from the combination of multiple mole melanoma syndrome. All the vertebrates, some of the invertebrates went blind. Even when there was still electricity, I remember a documentary on television that showed a desperate migration in the Serengeti that was simply too hard to watch. Same in Yellowstone. During the last days of TV. Total chaos. They either ran in the wrong direction or limped to their knowing deaths. I found a blind fly one day in my house. Strangest thing. Very sad."

"Hans, it was horrible. We all know," lamented Sophie.

"He doesn't know," Hans insisted.

"Actually, I do," admitted Petrus.

"So, even in Antarctica."

"Of course, especially in Antarctica."

"You saw with the birds, the penguins?"

"Yes, countless species. The birds, they still have no idea which way to go. I lost some good friends. I'm speaking of penguins, albatross, seals, and others."

"I am a cosmologist by training, and a generalist when it comes to all things Darwin and Ernst Mayr. So what do I care, you may ask," Hans said, calibrating Petrus' nerve endings. "You can't help but feel for a blind Dodo."

"Precisely," said Sophie.

"But in my journey, I have seen billions of insects and animals. All kinds of animals?" Petrus added, not by way of any objection to the theory of entropy, but from enlightened surprise.

© The Author(s), under exclusive license to Springer Nature Switzerland AG 2022
M. C. Tobias, *The Maiden Voyage of Petrus van Stijn*,
https://doi.org/10.1007/978-3-030-97683-5_28

"Oh yes, multitudes have prospered, hybridized, extinct creatures once, now in pleasant riot. No competition for niches. And no predation. If their genes have accommodated the UV, then they are assured of a long life, and many fit that category."

"And 25,000 people. Where are they, mostly? What category do we fit?"

"That's the question. But nobody really knows. The megacities have all gone the way of Paris," Hans said. "From what we could glean over the decades, am I right?" he asked Sophie.

"Yes," she applied nostalgically. "Speaking as a lucky one, I have to confess it all adds up to a most congenial population crash."

"And that's just largely common-sense extrapolations from the few cities here in Western Europe that have essentially been counted. By that, I mean literally every-one remaining in the town who wants to show up, comes to be registered at the central square. In Paris they assembled one June morning at the Place de la Concorde. In Brussels at the Grand Plaza. They say Moscow is a morgue, and India, China, Indonesia, most of Africa, essentially empty. Maybe on the remote island chains they had a chance. But only if there were mountains. The Caribbean, only a few of the islands would provide for people. Those gathered on Duarte Peak in the Dominican, perhaps. You must have seen it. The rising tides? I only know because, once, there was an observatory up there."

Petrus listened with mixed remorse. Not for any other reason than that he had never been part of these elusive, far-off congeries. These abstract disasters. He could not entirely get into the head, the emotions of one who had. But still, he had com-mentary to lend.

"Of course, we saw the tides. And the tsunamis. Triggered by unimaginable seis-mic tumult. By over 1,600 ft, on average throughout the Antarctic."

"1,600? Amazing. As of when?"

"By the time I was a teenager."

"That's remarkable," Hans braved, thoughts churning around as he tried to match that piece of information with the world's overall geography. "So, one can imagine Manhattan, Shanghai. But nothing like 1,600 feet in our own Ports of Zeebrugge or Antwerp. Water levels must be uneven, which would support my theory of the eerie oblong axis movement."

Petrus did not follow, but Hans continued: "So, finished, *finished*! but vastly circumscribed." He was wrestling in his head with the calculations. The puzzle of north Atlantic ocean conveyor belts, the demise of the Greenland ice shelf, upwell-ing from the great South.

"People fled," Sophie stated matter-of-factly. "Nothing else to do."

"Only to encounter waves of others fleeing," Hans added. "And fleeing meant being outside. On foot. The solar storms, having destroyed all electrical grids, meant, according to some, the end of the world. But that was just nonsense. What they failed to recognize was a profound, how shall I describe it – perhaps you have already sensed it – a wonderful ecological liberation. Maybe that's not the right description. I don't know. But we're not cavemen, as you can see for yourself. Not yet. We still have every mansion, coffee house, bookstore, art museum that we ever had. Not along any coastlines, of course. At performing art centers, the singers just have to speak up. As for nighttime events, well, the villages are lit up no differently than in Rembrandt's time. So, ebb and flow."

Said Sophie, "We should be, we are grateful. Indeed, more than grateful. There is no more threat of a nuclear war. Of aerial warfare. No gun has been manufactured in half-a-century, of that I am quite sure. Not a single vehicle on the roads. No air pollution. The ground water, however, requires great care. So many corpses. Two, maybe three decades of pyres. Smoke plumes covering the earth. You walked here from Gibraltar. You've already seen some of it, then. The flooded inlets. The cliffs along the coastline. Towns totally washed away, or, depending, covered in new forest. Insects, plants, animals never seen in Western Europe, and, of course, the animals they brought back prior to the crisis. What did they call it – de-extinction or something." She looked to Hans.

"Rewilding," Petrus added.

"That's the one. *Rewilding*. Well earth's Solar System made it all rather simple, I might add. Are there still all those apes in Gibraltar?"

"Yes."

"How many people there?"

"Less than two dozen or so, I guess."

"What else did you see, walking all the way here?"

"Every flower. Bush. Songbird. I studied biology all my life, from my mom, and in countless moss-filled cracks upon nunataks. That's a rocky peak sticking out of a glacier. It's very different. Yes, penguins and seals and whales. But not this, this mainland paradise. And everyone I've met has been so kind."

"Paradise?" Sophie echoed, partly to mock the notion, but also not entirely disagreeing.

"I wouldn't go romanticizing too much," Hans added.,

"Maybe best not to cast a shade on his bubble," Sophie suggested.

Hans clarified: "Just remember, human natures remain quite fickle. Even here, in the most beautiful city in the world."

"Is it?" Petrus enquired.

"Oh yes," Hans affirmed.

"And all these ancient houses, empty. Is that possible?"

"Why not? Of course, they are full of memories, possessions. But yes, unoccupied per se," Sophie chimed in. "If someone is residing in a house, they leave a traditional piece of Bruges white linen visible in a window, like monks in the old days. Of course, the art of Bruges lace making is over. Too hard on old arthritic hands. Anyway, you'll observe few such signs. Tell him about the castle," Sophie nudged Hans.

"Right. What you have not seen, young Petrus, is your own house."

"My house?"

Hans was tickled just imagining what was to come: "Oh it's quite impressive, I must warn you. I've kept the housekeeper, Neletje Girard, once a week just to dust and check for littler critters inside. I'll make sure she's there working first thing in the morning. The same with Rembertus and Dom, superb gardeners. Your father left you everything in his estate."

"What kind of estate?"

"By any standard, it was substantial"

"Oh go on. It's the nicest damned castle in Flanders," Sophie posited for the record

"It is impressive, embarrassingly so. Anyway, plenty of candles stored, along with the wine in the basement. Fortunately, the house sits on one of the highest points in all the town. So there has been no flooding on the property. Not so far from the old gardens at Zeeweg 147. You wouldn't know it. Just to be clear in my own head: this is your *first* time in Belgium, right?"

"First time anywhere, off the ice, that is"

"I can't begin to imagine. Amazing. Well, the entire region of Sain Andries has completely grown back, particularly to the north and northwest. Your father's castle sits now on the edge of hundreds-of-thousands of hectares of forest. Importunate grove upon glen. Lots of surprises for you"

Then, in a rapid turn of thought, Hans inquired, "Osna. What happened to her? You don't mind my asking?"

Yes, he had feared to press on details but really felt a need to know. The last time he'd seen his sister-in-law, Osna would have been in her early 20s. An intellectual prodigy, he remembered her well. She was obstinate, skinny, magnificently outspoken, driven in ways no one, probably not even Stefanus ever managed to fully appreciate, it seemed with hindsight. And with looks whose beauty was almost savage, undermining every other person in a room. He had seen the deterioration in his brother because of her absence, which grew increasingly desperate as the times closed in on everyone. The art curator had grasped with the lens of a moonscape that he had missed the entire growing up of his one child, this golden boy now sitting beside his uncle. And then, the protracted silence, until this day.

"You do bear a striking resemblance but, honestly, I was afraid to mention it."

"It's alright," Petrus said, almost dreamily.

"I'm really sorry."

"We were on a mountaineering expedition, seeking out rare biological specimens, during the worst of the storms. On our return to base, half the continent collapsed."

Hans was silent, absorbing the very sound of the words. Impossible to imagine.

Presently, the pelting hail just outside had suddenly abated, five seasons in 20 min, and Hans reckoned this was the moment to head out towards the house, a mile or so away.

"It was a pleasure, Sophie," all three stepping out the door of the cafe. "You'll see her again," Hans said.

As they parted ways, Petrus mentioned to Hans, "I did see a Dodo in northern France. With others of its kind. None of them were blind."

"That's good. They've become quite common here as well," Hans said, with a consoling grin. "They like to hang with the ducks. And harriers that are as much herbivorous as everyone else. Mind you, I know nothing about birds. Except that some fly, some don't."

Chapter 29
A Stroll Through Town

"The Square in Bruges," 1696, by Jan Baptist van Meunincxhove (1620-25-1703/04), oil painting, Collection Musea Brugge, Groeningemuseum, Public Domain

© The Author(s), under exclusive license to Springer Nature Switzerland AG 2022
M. C. Tobias, *The Maiden Voyage of Petrus van Stijn*,
https://doi.org/10.1007/978-3-030-97683-5_29

Hans led his nephew through the Renaissance maze of an emptied village, now mostly surrounded by water and forest. The historic neighborhoods had dropped several feet in elevation through the subsidence, leaving some houses tilted, sinking or completely corrupted, there having been no repair of infrastructure once the electricity was finished, Hans explained. Whereas some of the streets, and the town center, remained unaltered. Streets and houses six centuries old. A few still lived in.

And all of Bruges' magic remained, sans tourism, sans people in general. That was the point.

"As I said, almost every house or building is empty. Anyone can take up occupation," he elucidated as they strode past the Begijnhof, through the Koning Albert 1 Park.

"How does the economics of that work?"

"There is no more economics. And we're glad to be done with it, I assure you."

"But what currency is used?"

"Petrus, those days are over. I do give your father's former employees gold coins every week. They seem to think that someday such things may prove useful. But frankly *someday* has come and gone."

"What, people just trade things?"

"Essentially."

"And it works?"

"Well, it's quite arbitrary, your kingdom for a horse, as it were, but by and by, yes."

They kept on walking secret little lanes in and out of forest, over curving bridges, between closely aligned churches, canal houses and shops, mostly pristine, unoccupied.

"It really is beautiful," Petrus expressed.

"Surely your Antarctica is more beautiful."

"Different," Petrus opined. "You can see forever, but you don't see this."

"Tomorrow, once you've begun to settle in, I'll take you to what was always your father's pride and joy. The main museum."

Every tree, thousands upon thousands, seemed to be in full bloom. It was a cornucopia of cobalt, sanguine coquelicot and malachite hues with no natural wavelength, all throwing seeds to the wind, beneath a sky of charcoal storm and weather shifts. A warm Coleridge Xanadu mist was enveloping the puzzle of a park, as if by wanton stealth. The whole of Bruges was a sanctuary that seemed to extend forever. Goslings were furiously paddling to stay in their rows and keep pace with their parents through the canals, along with coots, red-breasted mergansers, ruddy and white-headed ducks, mute swans, pochards, mallards, and green-winged teals. Storks roosted in every soot thick chimney. Expansive flocks of snow geese flew yelping overhead and every tree appeared to play host to thrushes, titmice, thrashers, buntings, Old World sparrows and magpies. Pink-footed geese wandered beside pheasants and grouse. Turtle doves and plovers, waders and skimmers, gulls and loons could be seen in every misdirection. It was the feathery near end of a century, of a continent given back to its birds. And throughout it all, a Kingdom of Auks and Dodos. Kakapo and Condors.

Insects buzzed through the skies as if it were Suriname by night. Then, within minutes of Hans and Petrus' stroll, the sky turned pitch black, the color of that tornado prelude Gloucester sage, and a punishing hail slammed down in unrelenting vertical streams that broke off blossoms, turned vermillion pavers icy gold, sent thousands of birds seeking for cover. There was ball lightning racing like a jackrabbit from quadrant to corner, and huge rolling thunder throughout all of Europe.

"This is normal," Hans said. "Don't worry. Tornadoes are rare."

"So *many* birds." Petrus, who was no stranger to penguin colonies exceeding 100,000 individuals when he was young, was simply astonished.

"Many things," Hans said with the advantage of having been there, seen it all.

At some length they crossed an ancient but sturdy enough bridge and entered what had clearly been a well-trodden pedestrian path through a city park. Suddenly, both men stopped. Not fifty meters away, a herd of large cloven-hoofed oxen were crossing along a forest verge.

"Polish Wisent?" Petrus whispered.

"Aurochs," Hans replied.

"That's remarkable."

"Like I said, there are surprises in store for you. Dire wolves, as well. Perhaps your mother had something to do with it. It was, as I understand it, a relatively small community of scientists working on that kind of genetics."

"She focused largely on little things. Moss, lichens, although she also was part of an international group, in the beginning, all of whom were re-designing molecular structures that would have applied up and down every food chain. But there is a huge question here."

"Go on?"

"If there are aurochs, what about their contemporary predators?"

"Unclear," warned Hans. "In all the wisdom of the decade that saw so much experimentation, the rewilders were persuaded of the *red fox gene* [1]. I think biologists were surprised that the, what do you call it, Allee Effect [2], became an interspecies cooperative mechanism. Fitness, diversity, population sizes, correlations I don't understand, but certainly can see it for myself. But you would know more about such things than an idle stargazer. I have my doubts, I must tell you."

"Doubts?"

"I wonder how far correlations go. I'm speaking of those predators you referred to."

References

1. See Kukekova, A.V., Johnson, J.L., Xiang, X. et al. Red fox genome assembly identifies genomic regions associated with tame and aggressive behaviours. Nat Ecol Evol **2**, 1479–1491 (2018). doi:https://doi.org/10.1038/s41559-018-0611-6

2. *Courchamp, Franck; Angulo, Elena; Rivalan, Philippe; Hall, Richard J.; Signoret, Laetitia; Bull, Leigh; Meinard, Yves (2006-11-28).* "Rarity Value and Species Extinction: The Anthropogenic Allee Effect". *PLOS Biology. 4 (12): e415.* doi:https://doi.org/10.1371/journal.pbio.0040415. ISSN 1545-7885. PMC 1661683. PMID 17132047. *See also,* Roques L, Garnier J, Hamel F, Klein EK (2012). "Allee effect promotes diversity in traveling waves of colonization". Proceedings of the National Academy of Sciences of the USA. **109** (23): 8828–33.   Bibcode:2012PNAS..109.8828R.   doi:https://doi.org/10.1073/pnas.1201695109. PMC 3384151. PMID 22611189.

Chapter 30
The Castle

"In Bruges," Photo © By M.C. Tobias

© The Author(s), under exclusive license to Springer Nature Switzerland AG 2022　　107
M. C. Tobias, *The Maiden Voyage of Petrus van Stijn*,
https://doi.org/10.1007/978-3-030-97683-5_30

As they walked across town, Hans began, "I certainly don't claim to know much. But we all have lived with the results. Big, happy pussy cats, or 900-pound saber tooths, if you will. And yes, the woolly mammoths as well, their DNA extracted over half-century ago from remains on Russia's Wrangell Island. That was apparently a big deal."

"I don't know what to say. Good or bad. I don't want to judge my mom. She believed what she was doing was for the best."

"Quite a legacy."

Within a matter of minutes, the castle could be seen at the rear of extensive wild garden paths illuminated by centenarian sweet chestnuts, limes, and Giant sequoia. There were Oriental plane trees, even cedars of Lebanon, oak, and spruce throughout the unmanicured arcadia. The fountains were all dormant. Giant hares darted here and there, while enormous Korean Crested Shelduck, Mascarene Elephant Birds, and Birds of Paradise helped themselves to berries and tubers, grazing on the sweet grasses. A Molokai 'O'O darted betwixt soft branches of a rare Three Kings Kaikomako, from the southern hemisphere. The 'O'O's tail feathers were serrated not in white but silver, its honeycreeper beak the color of a violet sky, a mutational flavor of the month, perhaps. And beside them, the incredible Rodrigues Solitary. The 16th/seventeenth century Dutch fabulist painter and documentarian, Roelandt Savery could not have imagined a more perfect harmonium.

"It's all yours, Petrus. I mean, if you want it. Your father spent many happy years here. Despite the chaos of the times, he found safety and sanity. We spent countless joyous days and nights together, here."

"It's too amazing." Petrus gazed upon the five-story castle built of a regional bluestone, the Petit Granit-Pierre Bleue de Belgique from Soignies, Hans explained. "The house has stood for many centuries. But after NATO was abolished, and there were those incursions from Russia, much had to be rebuilt, and some of the rooms were left as attics. Still, I think there is enough space for one person."

"There was a war?" Petrus asked.

"Not a real war, that is to say, an end of days war. Pipelines were detonated. Tens-of-thousands gunned down. Fierce fighting. But it was all, shall we say, normal outbreak of deeply rooted enmities about power and water. But as you must know, all the nuclear armaments were disabled permanently by the geomagnetism. Thank God. Otherwise, well, unthinkable. The magnetic shifts were far more powerful, but benign, than those conventional EMPs associated with such a war."

"All I know is since my childhood there was never a signal, except between two bases on the ice, briefly. And, a distress code from here. Bruges. Then, only noise. After that, the static went white and silent."

"From Bruges. eh?"

"Yes."

"Well, I'm going to show you what that's all about, tomorrow. And hopefully, you'll be guided by reason."

"What do you mean?"

"You'll see."

As they approached the enormous entrance, a regal and ornate front door of mahogany and elegantly sculpted bronze greeted them.

Pointing to the workmanship, "Your father loved and could afford Bramante," Hans said - huge, heavy handles, meant to dissuade, as four poodles of varying size ran up to greet Hans and smell out the newcomer.

"That's Picasso, that's Sartre, and those are the van Eycks. All related. One day they just showed up."

"You named them?"

"Yes. In honor of your father. He knew them for less than a year."

"What do they eat?"

"Most anything, But, actually, again, the genetics."

"I don't follow."

"A hundred years ago, as I understand the basics, it took some twenty years for a wild plant to become a domesticated one. Then it took twenty minutes, and then ten seconds. In the famous case of the Dmitry Belyaev fox population it took all of two fox generations for them to become totally tamed. A Mongolian gerbil doubled its size in two generations. Mozambique female elephants required merely twelve years to evolve no tusks as a deterrent to poachers. Whale sharks apparently lost all their coloring within one generation to conceal themselves. The beak size of the Everglades snail kite grew within a matter of years to enable those birds to be able to eat the invasive larger apple snails. White throated Canadian sparrows changed their songs within months. A finch became a new sub-species in a matter of twenty years on the Galapagos. And so on and so on. But you probably know all that."

"Hans, we got very little information. If it wasn't happening in the Antarctic, we didn't hear of it."

"Well, let me tell you the good news. Feral cats no longer go after birds. Great white sharks are harmless. It appears that every product of directed evolution was programmed to be herbivorous. Again, I guess we owe some of that to your mother."

"What about red necks?" Petrus wondered aloud.

"Come again? A bird?"

"I mean ass holes. Is there a red fox gene for ass holes? We certainly never heard of it in Antarctica."

"Well, of course, fascists in country after country who made a stink. American Republicans, Austria's Black-Yellow Alliance, Vlaams Belang here in Flanders, individual liberty groups throughout Europe who claimed adherence to the Utrecht Declaration, a most rotten piece of paper, certain rurality movements, and the German Citizens in Rage. But interestingly, cancer winnowed populations so rapidly. The chemists had no problem overseeing most populist parties. The potions were pumped into the water supplies of every nation, long before the power ran out. Unthinkable, previously. I mean in any Democracy. Who would have ever thought that it was going to be skin cancer that ultimately unified remaining survivors. Geneticist used to call it the kindness gene. The demographic crash., coinciding with civil wars almost everywhere, selected for that kind of kindness, in the end, which we are now inhabiting. Like Munro Leaf's Ferdinand, you may have read it? The loving, flower child of bulls. And the selection broke open. I don't have the geneticist's lexicon to describe it, but it was unleashed in every direction. Animals, plants, everyone was affected by it."

"As a biologist it's hard to believe. But then, my mother promoted just such a vision."

"So did your father. Although he derived his ethical convictions from art, not genetics."

"It's hard to swallow, though I agree. I've seen what I've seen. And it appears to be the case. We had a great greenhouse at Princes Elisabeth Station and most of the occupants abstained from any kind of animal. It was frowned upon."

"You were cut off. From everything. Amazing. But, like I indicated, it's not smart to wax utopic. Those were very dark times in the beginning. And there are some very toxic places, still. It's not all a picture postcard. Three miles further on, across what is now your land, a forest of historic potholes, is a cemetery. At first it was an impromptu emergency burial site. Then the corpses multiplied. The fiery pyres could be seen burning day and night from miles away. Something out of Leopold the Second's Congo Free State. Belgium's greatest stain. The incinerator scrubbers everywhere clogged with too much ash. We have no clear idea how many actually died, but almost everyone did. Then Cholera and weird zoonotics followed."

"We saw the majority of occupants at our station die."

"What causes?"

"Everything. Then we lost all our primary power sources I think in the year 2037 or so. After that, it was emergency concoctions. Cobbled together, no power supply was ever again uninterrupted."

"2037?" The year seemed to impact Hans. "Are you sure?"

"I think so. Why?" Hans gestured that it didn't really matter. "At least that's what Osna told me. And even all the back-up generators. We went dark from that time on, save for the fuels in storage, and various technologies to maintain the batteries, stacks of fuel cells, electrolyzers. But there was an electrical point of no return. When all communications were lost. Not a volt on the entire continent."

"Isn't it interesting how, without compass, watch, light, power, the time stops fighting us, or we stop aggressing within the limits of known time. The distance between a casual, taken for granted basic, to a desperate essential is an inch, a nanosecond."

"I guess," Petrus said meekly, distracted by nearly everything.

"Here the year was 2031, but I can't frankly recall if that is adj or pre-adj. But you're saying you guys had access to a satellite until 2037?"

"Really, Hans. I'm not certain."

"Well, it's interesting. Here, with the end of electricity, we simply lost interest in time, is what I'm trying to say. It was not like fighting for breath. No one felt the need to fight for seconds or minutes. Life instantly became much easier. The garderobe, probably four hundred years old, might surprise you."

"What's that?"

"Toilet. Sorry. It's a modern one, no medieval privy, in the master bedroom, a direct discharge into the outdoor cesspits."

"And drinking water?"

"There is a reverse osmosis system, but the filters are cotton. It's imperfect. The water comes from a spring in the forest. Your father never doubted the quality. But there's no way for us to check it. I think it tastes fine, a bit hard. There's a dedicated wood burning stove and quite an ingenious system for heating water for a bath, pipes throughout the house. This is not your normal castle."

They entered the dark, humid interior. Hans lit numerous oil lamps. "No dearth of traditional light sources," he added.

Petrus gazed at the vastness of his late father's universe. So many books, paintings, tapestries, antiques.

"It's dusty. Most of his art collection is in glass. The moths are always a problem. But, honestly, it's all okay. For years I've come on the weekend to look after things, though I prefer my house in town. Anyway, there are a dozen bedrooms to choose from. Let me show you my favorite. It was where your dad preferred to sleep."

They climbed the elaborate, carpeted stairway to the second floor, then trekked down the curving hallway to the end chamber, floor-to-ceiling-stained glass windows overlooking forest as far as every horizon could be deciphered, through the receding cumuli of another storm building up.

"Not too shabby, don't you think?"

"Northwest," Petrus at once determined.

"Yes, 19 miles to Port Napoleon, which is nothing now. Brussels, Amsterdam, Antwerp, so few people. It is quite striking. You are best right here."

"The forest is fantastic."

"A squirrel could move from here to Siberia without touching the ground, at least that's my guess."

Petrus wandered throughout the room. "Quite a gallery."

"Portrait of a Young Girl," c. 1470, by Petrus Christus (1410–1475), Gemäldegalerie, Berlin,
Public Domain

"Many of your father's favorites: Rogier van der Weyden, Lucas Van Leyden, Petrus Christus – that's probably where your name was taken from, gorgeous, isn't she? Leonardo's Ginevra couldn't touch a candle to her. And here, a Giorgione, Hugo Van Der Goes, and Gerard David."

He drew close to the Petrus Christus.

"You can take it off the wall. Go ahead. Hold it. No security. And it's yours, after all. She looks familiar, no?"

Petrus saw at once the strange likeness to his mother, when she was very young.

"I imagine it's a lot to take in," Hans reckoned.

"How did he afford all this? I mean, they're expensive, right?"

"You obviously never knew your grandfather."

"No. Only a few stories. Some lost photographs."

"He had inherited millions of shares in what was to become the largest mustard manufacturer in Europe. I must say, it's good mustard. You'll find enough jars in the wine cellar to last a lifetime."

"Mustard paid for all this?"

"And avocados, from Central America. Your grandfather owned thousands of acres of them. Before the cartels got involved. As for Stefanus, he remained a passive investor in agriculture, including funding for what was called, I believe, C_4 anatomy. Biophysicists who tried desperately to improve upon shall we say *conventional* photosynthesis. Maybe you know that. Or Osna did. Conversely, Stefanus was certainly no slouch when it came to buying and selling art, in his spare time. He dealt privately despite his full-time senior position at our local Groeninge. The Museum. That was his true home; where he was senior curator for decades. Nothing has changed much over the years, except, of course, for the lighting. Your father knew paintings. Sometimes I thought he was able to look through the many chemical layers of a canvass or piece of painted oak, deep beneath the surface, peering with an unabashed voyeurism through the imprimatura, the one multi-dimensional maze of craquelure, and learning everything about the artist's personal life. For him, it was like reading a book. Or as I might look for new stars through the best available telescopes. He could gaze directly into the souls of the early Flemish artists. And decipher what he took to be their secret languages. Codes."

"What sorts of codes?"

"Just like that? No, Petrus. In your own time. My answers are found in a hot chocolate. A clear night in the middle of the Milky Way. Questions and answers will be different. I've had many years to do nothing but think. And eat. And sleep. But I will tell you that your father was certain that the artists he so loved were entirely predictive. He believed that they painted the future. Was he right? Very likely. Bosch painted hell, but also paradise. I will tell you that your father and I were always close. We had some trivial differences, but he was my older brother and always looked after me. Our views on so many things were aligned, from the opening stanzas of Dante's *Paradiso*. Have you read it?"

"Never."

"Lucky you. Now, you'll have all the time in the world to do so. Stefanus was not so much a religious man but one fully captured by the arts. He believed in Dante's opening line, *Glorious, that Being who animates the universe and is all aglow*, or words to that effect. But then Dante went on talking about that forest. Yes, that one right outside. A divine realm, enshrined forever in his heart, in your heart, he hoped, beneath the very trees we love, and like a tiara of green leaves upon our minds, though memory cannot follow us – I'm slightly paraphrasing, of course – cannot follow, but is fallow, pregnant with life and that is the most worthy of all themes. There are copies in the library. You'll want to read everything. But he had one particular wish, toward the end, when the cancer had spread so quickly. He prayed that you and your mother would come home and carry on. I suppose he realized that while it was too late for him, there would be enough time for you." Hans paused, trying to figure out how best to put it. "All parents do."

"Time for what?" Petrus asked anxiously.

"For life. To survive rather than die. That's all. Just to survive and to try and enjoy what's left of it."

Chapter 31
Acclimation

That night, Petrus was left alone in the sprawling castle. Hans had shown him every-thing necessary, and, after a quiet dinner, wandered back along dark, silent paths, then cobblestone streets, to his own home in the Genthof district, a thirteenth cen-tury, five story former manse looking out across the Gouden Handrei canal.

Petrus patiently moved from room to room in the estate, holding a lantern lit with a flat cotton wick. He used matches from a bar known as the Café Vlissinghe, which meant nothing to Petrus, though he later learned it was the oldest such establishment in the entire town, dating to the early 1500s.

But for now, and just like that, his attention was glued to a not so far-off animal call. The smiling turquoise whale, *Balaenoptera olli subredens mysticedi*? he imag-ined. It was a species that had only been discovered in Antarctic waters in 2041, and was larger by more than 10%, on average, than blue whales. The Belgian coastline was close by and it was certainly possible. The blue whale's 17-Hz frequency was the equivalent of nearly 90 meters of length in the water, a sound wave that could travel the distance from Paris to Brussels without any difficulty in times of no known radio interference. But this cry felt earth-bound, with a terrestrial edge, like some vast groaning question, equal measures of deliberate calling out and a wall of static. But it emanated from the forest, not the English Channel.

The kasteel or burcht, or whatever one called it, extended to ridiculous extent. At least 40,000 square feet, Petrus calculated. It would take a mouse many days and nights, even under a free reign, to account for all the opportunities. A spider could live forever in this place. A mite for multiple eternities. A housefly could savor its 2 weeks of life, exploring every surface without surfeit. No one could stay on top of so much dust. There must have been three dozen large wrought iron chandeliers throughout, most of the candles burnt out. The rugs were in very ragged condition. Books, including many very rare ones by Sannazaro, Vasari, Goethe, Petrus at once recognized, had been moved about, the pages revealing foxing, the bindings bearing the time-worn wormholes. None of the paintings seemed precisely in place. The elaborate edifice was just that, an empty mansion, filled with lonely memories.

© The Author(s), under exclusive license to Springer Nature Switzerland AG 2022
M. C. Tobias, *The Maiden Voyage of Petrus van Stijn*,
https://doi.org/10.1007/978-3-030-97683-5_31

Petrus searched for any layer of personal recognition. He had evolved from polar bunker to desperate iceberg and onward to land, a postdiluvian cage of memories liberated from their silo into the Biblical garden, absent any apparent finalities.

A badger, possum, or beaver scooted down a hallway. Petrus lost it in the darkness, beguiled by the few seconds he managed to see its ravishing pulchritude. The sound of rats, probably the ship or Norwegian ones, chewing methodically and delightedly upon wood or metal pipes, slate shingles or ceramic, was manifest in more than a few rooms. Why not? And what better place? Perhaps they were squirrels, or even magpies or raccoons. They had their paradise.

But it was the photo gallery in the master bedroom that most interested him. There were his young parents, probably not long after they had first met. They were enjoying themselves with friends in some restaurant with low, Medieval-like ceiling beams, probably here in Bruges. And there were images of his grandparents standing awkwardly before some wild sculpture of tulips. He had no idea, previously, what they looked like. Stiff enough portraitures to fall over in an earthquake. Self-conscious. Their thoughts unreadable. Both were dressed in formal attire, his grandmother in heels. He discovered a closet in a back room down the third story musty hall, with varicolored chemises still on hangers, dozens of them.

He shook out the bed sheets, ignoring the odd rodent turds, then refitted them, beneath a heavy set of antique blankets, the fringes entirely gone, eaten through by millers. Nothing seemed to date from the twenty-first century, even the bottle of 161 proof rum, from somewhere in the Caribbean, on a side table, adjoining a little framed pastel, not exactly calming, by one Edvard Munch he'd never heard of. Petrus removed it to an out-of-the-way shelf. Then meditatively supped on the exotic libation.

The bed was a double king, the size of three algae tanks back at pes. Sizeable enough on which to rehab a couple of Weddell Seals. Petrus felt embarrassingly fortunate that night, but quickly passed out before the sensation might have driven him outside, into a nearby barn, where his comfort level, his very instincts lay.

He dreamt that night, as so many earlier nocturnal fests while coursing the southern hemisphere on that floating ice plateau the size of Sikkim; imagined some kind of twisted world where matter and antimatter were in constant debate as to why life should ever arise. What good was it?

In his dream there was something like a coagulating, tormented DNA molecule, its frenetic polynucleotide chains angrily wrapping around a city in which mobs of people struggled to make progress through blistering multitudes. Everyone was angry, demented, on opposing sides of every issue. But therein, no one but Petrus could be found at the other end of the braided thing called life. He had no pockets, no possessions, nowhere to go, only to flee.

It was only a peculiar, if informed dream. One of a myriad. Always emphatically stamping the same impulse of incoherence: human natures run awry.

There did not seem to be any purpose to the stormy fragment, this mental outlier. People reacted from frantic impulse. And then the dream would transmogrify to all those he had buried in the bitter gales around pes. Somewhere out there, her organs ruptured, bones shattered, spilled blood congealing rapidly. Osna. Stolidity in the massive cold. For years she had put aside any thought of regrets. That there would be no way of ever returning to the green world north of the icy continent. Even wondering if it in fact existed anymore. She rarely spoke of these matters. Other than that near mystical, frightening, arhythmical signal, electronically inexplicable, coming from Dijver 12, 8000 Brugge, Bruges.

He would awake in storm, or inscrutable dense fog, or beneath starry countenance and realize that he and perhaps a few remaining stoical emperor penguins were all that was left of the life he had lived, happily and without the least lament, for three decades. Only to watch the last of the birds drop into the icy seas, no sentimental farewells, and then the whole last remnant of the Princess Elisabeth Station disappear in an instant and plunge into the Atlantic. Were they warm waters, cold waters? It didn't matter. All that information he and his cohorts had cultivated for decades, was of no importance now.

When he awoke next morning, Petrus could not determine whether he was smothering beneath a great unshakeable alien weight, a terrifying surmise with no obvious cure, or if, by simply looking out the leaded beveled windows upon miles of forest, rather than of unrelenting icesheet, such teeming life had lent to him the ultimate reprieve. Let no man doubt his happiness this day. Mesmerizing and ferocious though it was.

He had first touched olive and carob trees in Gibraltar, marveling at the pods, the scents, the crowded veins in the leaves. But the real revelations emerged as he trekked through Los Alcornocales, with its ancient labyrinthine cork oak species, holly and bay laurels, and rhododendrons in thick bloom amid an Eden of mosses and little pocketed waterfalls. And biomass flying everywhere. Prior to that time, only in the greenhouse at pes had he known plants any larger than the native Antarctic grass species; bonsais, a miniature pear tree, bush tomato, radishes, cucumber and potato, parsley, and cilantro.

Presently, wearing his father's thick cotton pjs and bathrobe with SVS embroidered on them, he gazed upon the thousands of acres of forest. Native, invasive, a sylvan chaos of lush ecodrama that rose up in so voluminous an audible medley as to mimic the greatest symphony halls. With time he would come to know the silver maple, Southern catalpa, the London plane and Black poplar. To measure by hand the girth of Bald cypress and Giant sequoia. His fingers upon the barks betrayed a principle he had missed thus far in life. To feel without the clouding of facts. That omnipresent pulsing reality, half-imagined fairies running up and down the trunk, each leaf an evotranspiring miracle unknown to the last continent, but now gushing in upon the all vulnerable heart of the new world with its infinite occupants, cell messengers of every biological esprit de corps combating any notion of a personal or global void for foundlings. The endless living connections were spellbinding.

But however revelatory these fine tidings; no matter every ounce of living burst and strange, perfect, novel companions, he first had to figure out the most basic of self-perspectives. Questions posed by Buddha, Plato, Gauguin. Something that cannot be solved.

Petrus now sensed that every organism, species, individual, Kingdom of life, asked a single question and submitted to the evolutionary conundrums, bimatrix games eclipsed by an orientation of endless probability. Taxonomy was no longer relevant. Only an individual.

What am I to do?

Chapter 32
Basics

The next morning, Hans knocked on the castle's front door. It was a large bronze knocker, a sculpted Shakespearian *fool*, taken in a previous century from some church whose congregation clearly appreciated the ironies of life.

"Hello? Petrus?" he called out.

Petrus and the dogs were out back exploring the property, Neletje, coming to the inner courtyard of the estate to greet him, explained. And Rembertus and Dom were in the greenhouse.

Hans came back in the afternoon and this time found Petrus seated at the enormous carved walnut dining table across from Neletje who was standing, a mop in her hands. She was a hard looking woman probably in her early 70s. Other than a trip as a child to Brussels and Paris, she'd been nowhere, Hans had indicated.

Neletje had worked for Stefanus for many years. Her children had died. Her husband had wandered away, diseased, to perish somewhere in an estuary at high tide. His body was honored with indifference. Today, there was little obvious loyalty to the newcomer. No reason but a general bitterness, Petrus felt. For all that, like Rembertus and Dom, it was clear she appreciated having something to do. To relieve her mind of memories or thoughts of failure.

Light shone in from literally hundreds of windows of various sizes. All of them had off-white louvred shutters of lime wood, and solar shades on the inside that one could easily manipulate with cords of frayed twisted local fiber.

"I came in the morning but you were outside somewhere," Hans began.

"Yes. I can't quite grasp the forests, not yet. But have enjoyed getting to know Ms. Girard."

"Neletje," she said, forcing a smile.

"Good to see you smile, Nel," Hans said, by way of wise advice. Then asked Petrus, "You feel up to a stroll into town?"

"Absolutely."

"How did you sleep, by the way?"

"Not much. I don't remember."

© The Author(s), under exclusive license to Springer Nature Switzerland AG 2022
M. C. Tobias, *The Maiden Voyage of Petrus van Stijn*,
https://doi.org/10.1007/978-3-030-97683-5_32

"Creaking roofs?"

"Yes. That. And the rats."

"Rat and bats, squirrel ferrets and bobcoons. The house dates back originally to the thirteenth century. There have been one or two repairs over the years, of course. A storm in the thirties shattered all three hundred or so windows. Lightning strikes, a short-lived civil war, a fire caused by the funeral pyres that got out of control. Here, wear these."

He'd withdrawn from a shirt pocket a pair of thicker sunglasses than those Petrus had been utilizing, UV 2200.

As they entered the old city there was quite strikingly almost no one to be seen.

"A ghost town," Petrus mused aloud.

"In the town proper, fewer than thirty-five or so happy souls, as I'd mentioned." Hans thought about it, then bespoke, "I always preferred the sterile birth canal of stars to earthly biologies."

"Not a fan of life?"

"Just slightly misanthropic. Too many disappointments, I suppose. Oh come on. I'm just venting. Of course, I love life. I've had years to read up on the genetics, the biology 101 we physicists usually ignore. I ran out of physics. Mathematics. It will be interesting, like some case study, to see your reactions to all this, here, Bruges. A life in Antarctica, and only Antarctica, well great galactic viewing and the penguins must be of remarkable persuasion, but I can't imagine it. Nor could your father."

"I'm homesick for it. I can't deny that. It was simpler."

"I've heard something similar from a former prisoner. Trust me, you'll soon come to appreciate a fine wine and chocolate. All the great art."

"I'm quickly developing a taste for it."

"Look, I am no absolutist. But I could live without hops if it meant no Hobbes. I had two children, your cousins. Both died young. The brutish and short aspects of nature one never gets over."

"I'm so sorry."

A brief silence between them.

Then, "What were their names?"

"Virginie and Maurice. The cancers took both of them mid-life. She was a painter, he, a linguist. You would have liked them."

"The thought of cousins, I must say, feels strange."

"They weren't strange. Just victims. So, that's being human. Being a father. Both my children I can say were dear friends. That's not always the case. Let's not talk about it. Better to show you the most densely populated building in all of Brugge. The great art you mentioned. The ghosts still live there, quite enviably, I would add."

First Hans led his nephew to the house at Gouden-Handstraat 6.

"What's here?" Petrus asked.

"Nothing really, now," said Hans. "But for his final twelve years, Jan an Eyck resided inside, with his wife Margaret, until his death in 1441. Your father was convinced that Van Eyck, probably the greatest of all painters in history, derived inspiration for many elements of his landscapes from the forests and what was once a large pond behind the castle, your house. And Stefanus always said that there were missing pages from the famed Turin-Milan illuminated manuscript. It was illustrated not by Hand G, so-called, but definitively van Eyck himself, hidden somewhere in the castle, but he never found them. Do you know about the van Eyck brothers?" [1].

"Only the name," said Petrus.

"Then you are like a child about to look at the universe through a telescope for the first time. In fact, Jan van Eyck may be the only artist in history to whose work you could apply a microscope and not be disappointed."

"A microscope's not always necessary," Petrus added.

"That is true. Perhaps, especially true. Same with the telescope. These days, I much prefer just looking on my own. Your father and I used to talk about colors until late in the night. The naked eye can see S Monocerotis, for example. The astonishing Christmas Tree Cluster in NGC 2264 [2]. Everything about it, magnitude, variability and so on is simply the gobbledygook of an old astronomer in love with the bright colors of the Universe. Your father used a little telescope I'd given him. I have no idea what happened to it. He used to run star temperatures and the common spectral lines by me. I tired eventually of teaching him about that stuff. Just look up and enjoy, I'd say. He had this great eye for color, you see, but he complained of a stiff neck. 'I prefer looking right in front of me,' he concluded, succumbing to the colors of Brugge."

"I loved looking at the stars from pes."

"That was always a dream of mine. To see space from Dome A."

"You know Dome A, Yes?" Petrus asked. "That was the Chinese Kunlun Station. It's the highest point on the ice cap, over 4,000 meters. Frighteningly cold. Gone now."

"Gone?"

"Blown away, literally, was what we heard."

They followed the once bustling lanes, along the Reie canals, from the Zand to the Bonifacius bridge, taking in the major sights. The Basilica of the Holy Blood and Church of our Lady. They wandered the empty Groenerei, Augustijnenrei, the house in the Pieter Pourbusstraat, and back around again. Past Sint-Salvatorskathedraal, where van Eyck was buried. And on to the Gruuthuse Hof, to Guido Gezelleplein and Arentshof. And there stood the Groening(e) Museum, rising from a long-neglected consortium of overgrown baked red brick walls amid a profusion of flowers and trees in bloom. The whole complex, Hans explained, was once owned by the Sisters of the Sint-Andreasinstituut on the place where the Eekhout Abbey was first dedicated to Saint Bartholomew in 1130. It all just passed by the newcomer. He felt stalled, a kind of blank map that resisted being re-filled. But in this profusion of life, he no longer missed the various cries of penguins, the gold and magenta topographies punctuated by wind, held hostage by conditions. But he was always in a hurry to reach the next mountain range. In retrospect, he presently could not remember why. With so much biological bounty filling in every conceivable gap. No mountain ranges needed.

"Here we are, Petrus," Hans quietly announced.

"It's very tranquil."

"You lived on the ice. He lived here. In fact, your father rarely missed a day when he didn't spend at least an hour or two decoding the works."

"When you say, decoding?"

"Yes, much like my thousands of hours down in Uccle, a municipality in Brussels, at the Royal Observatory. In the days when we still had electricity, we found numerous minor planets, and detailed countless anomalies with regard to black holes. All to no end, of course. Like continuing to debate the luminiferous aether and the propagation of light."

"And my father?"

"Stefanus found his own black holes. Though they were never actually black."

References

1. See *Hubert And Jan Van Eyck*, by Elisabeth Dhanens, Alpine Fine Arts Collection, Ltd., New York, NY, 1973.
2. See *Colours of the Stars*, by David Malin and Paul Murdin, Cambridge University Press, New York, NY, 1984.

Chapter 33
The Codex Stefanus

"The Virgin of Chancellor Rolin," 1435, by Jan van Eyck (1390–1441), Louvre Museum, Paris, Public Domain

© The Author(s), under exclusive license to Springer Nature Switzerland AG 2022
M. C. Tobias, *The Maiden Voyage of Petrus van Stijn*,
https://doi.org/10.1007/978-3-030-97683-5_33

No one was in attendance to guard what was the greatest collection of Flemish masters in the world. Even the Ghent Altarpiece, by the van Eyck brothers had been moved to the Groeninge museum when a fire predicated by the solar storms half-century before swept through Saint Bavo's Cathedral. Hans described how it was that miraculously Ghent locals were able to remove all twenty panels in time to save them and it was Stefanus van Stijn, chief curator in those days in Brugge (Bruges), who opened a fifth, fifteenth century room at the museum to hold them. They were carried by 20 draft horses the 33 mi to Bruges, the panels covered in thick leaded X-ray aprons.

"No one to protect the contents?" Petrus asked.

"Not needed," said Hans. "As I mentioned earlier, the rare survivor has pride but not avarice. Pride in Flemish history."

"No crime in Bruges?"

"None," Hans declared with certainty. "As I said, no economy, no money, no banking systems, no governance. People are free. And what does freedom mean?"

"Does it really even exist? Isn't that a matter of personal choice?"

"Then you've answered half the greatest debate in history. Now you will find out the other half. I think for hundreds of thousands of years we've all wondered. Today, like it or not, freedom is our condition. Not to wax philosophically, but it is a set of circumstances that clearly allows for free will, notwithstanding the fact the Universe is unaware of us and one of these days an asteroid will instantly annul the marriage that is life. I don't just believe that. I know it to be statistically unavoidable."

"Nice," Petrus said with a neutral acknowledgment.

"But enough of that. Just look where we are. Enjoy the moments."

Both men grew silent as they wandered from room to room, painting to painting, concluding with the last few additions from the early 21st century.

"What comes to your mind?" Hans asked at length.

"The sheer beauty. Obviously."

"Um-huh."

"And the history. Mostly gruesome, I would guess. Gaunt, stern, overwhelmed by the edicts of religious scruples. Top heavy obsessions with paying obeisance to Christ, to Mary, several Mary's; to pleasing God, I gather."

"All true."

"On the other hand, a definitive flare for fashion, of their time, that's evident. And, as you said last night, the vibrant colors. Time hasn't dulled them."

"All true. The chartered guilds were busy here. Sheep herders, woad-growers for the cochineal dye, indigo came two hundred years later. There were countless cloth-weavers, the dyers themselves all going after madder, scarlet, saffron lily and lac-mus, a German lichen, you must know of it. Now you've seen fabrics in every blue, particularly Beverly, or in Pan de Carleta, or yellows and greens. Tapestries insulat-ing every room. Colors gone wild in a prolix pantheon of drunken show. Candle-light and colored wool transforming interior glooms to resplendent views of bright and sunny worlds. But you're a biologist. Look harder," Hans matter-of-factly advised.

"For what?"

"Your father's alleged code. He was certain of it."

"Now you're messing with me."

"No. Seriously. Look more closely."

They reversed their itinerary and headed back to the first gallery. They meandered through one silent room after another. The low ceiling corners had been outfitted sometime before 2031 with a few heavily protected skylights, ordinarily out of the question in a museum, but these were strictly protection against the augmented UV. Certainly better than risking some ignoramus wandering through with a torch.

They passed along the "Triptych of Saint Erasmus" by Dieric Bouts, Erasmus, an early 4th century bishop of Antioch, being martyred in his white loincloth. The painting's rear landscape was no wilderness, like that of Joachim Patinir's "Landscape with the Flight into Egypt." Rather, a completely docile, domesticated, world that had succeeded in combating chaos. There were human and other animal pathways, few trees left standing. There was a Joos de Momper II landscape equally disbursed by human order. As if God had ordained a rage for the mundane.

But then, with the "Master of the Legend of the Holy Magdalen," a Triptych encompassing the Virgin and Child, as well as the Saints Catherine and Barbara, painted around 1500, and called a "landscape" Petrus gazed upon three of the most gorgeous women ever realized throughout the Renaissance. A recipe for thinking of nature as sheer magnificence in female form. All the trappings surrounding it were sublimated to their sublime presence.

Petrus was quiet, striving to take it all in, trying to feel what his father must have felt.

"Any age that could produce these three ladies defies any other verdict than one of utter sophistication," Petrus finally concluded aloud, suddenly thinking back to Dulce.

And then they moved onto Jan van Eyck's "Madonna with Canon Joris van der Paele," exquisitely executed in 1436.

"Those are the heraldic colors of Bruges," said Hans. "Jan van Eyck had five years left to live when he did this, four years after finishing his astonishing Ghent Altarpiece. Come."

There, in its own little room were erected all the panels of that masterpiece. Neither Christus nor Memling moved Petrus in this way. Yes, the others, like Rogier van der Weyden and Bouts, Gerard David and Jan Provoost, Massijs, and Gossaert exhibited their respective genius with its complements of impact. But these panels by the Van Eyck brothers were taking the art form to some entirely new level. Shocking, world-combining. This was the essence of biodiversity through the human lens. The *Mystic Lamb* could not be more moving, his eyes expressing all the grief, irony, bewildering pathos that could ever exist [1].

"It is amazing!" Petrus sighed.

"Any code?" Hans asked.

"Code." Petrus walked back and forth along the panels. Finally, "Well, of course there is the very manuscript of being, of the birth of life, of suffering and prayer, of joy and jubilation. Everything I can feel is there. This is no code but a codex, an entire Book, not unlike the Bible, but in it is the future realized. As if, right now, that is precisely where we are standing" [2].

"I don't disagree. I'm happy you feel that way. Your father would have been pleased to hear you speak like that."

"But there is no one to protect it?"

"As I've indicated, crime is no longer with us," Hans began. "Although probably everyone has a gun, and bullets to spare. From practice most of us keep them close at hand." He removed some very light-weight handgun made of plastic from an inside pocket of the vibrant blue cloak he was wearing. "You see? There are several sitting around, loaded, in the castle. Remind me to show you. But to answer your question, in reality, there are simply too few people, Petrus. Who would bother trying to walk away trundling such bulky treasures, though a few panels *were* stolen over the centuries. In different times. Now, we are fortunate to live in a world where there is no buying and selling on any market. That's all over. And how lucky we are because of it. No more need of Anselm or Locke; of a Mills, a Ruskin or Gandhi or Hugo Grotius. You see the point: no more endless ethical speculation. All those arduous days and nights of civilizations aspiring, prolonging and suffering, are enshrined here in Bruges, or Brugge, if you prefer, and have been for a few thousand years. Even up to this day, you and I, standing here having endured a near total human extinction level event. What's left, amazingly, is all good. There is no need to speculate or offer advise on how to fix anything. Such is the happy outcome of several decades of Apocalypse. Which is what half of the works in this museum labor over. All the coming plagues, wars, desolation, and whether there is human free will that can be brought into compliance with the wide ethical global view of a divinity. The Ghent Altarpiece, Jan and Hubert van Eyck, they are pages of the greatest anticipation ever envisioned. Now we are there. Yes, it is fair to call it what it is: paradise. You don't steal paradise if you're living in it."

At some length, they left the first gallery through a door concealed in a wall, which led to steps angling down an old stone cavern, as if heading toward a crypt illumined only by very old glow tape along molding junctions. There, they reached a room which required a rusted, oddly constructed key to open. That key hung like an inconspicuous cross on a little black necklace around Hans' neck. He fidgeted with the lock, he knew its secret, and they entered a studio. It was like that of St. Jerome's from the famed Dürer etching, absent a lion sitting there. The chamber was stuffed with books and strewn documents, and large table. "This was your father's office."

Petrus looked at the reference books dating back to Bartolomeo Facio's *De viris illustribus* published in 1456. To studies by Michiel, Summonte, de Guevara and the great Italian Giorgio Vasari. The earliest short biographical references to Low Country painters at a time when Italian artists were the rage. Dominicus Lampsonius' *Pictorum aliquot celebrium Germaniae inferioris effigies* of 1572, and the extensive *Het Schilderboeck* of 1604 by Karel van Mander.

Suddenly, Petrus asked, "What is that noise?" He recognized it.

"2031, at least as far as I am aware. You indicated it was six years later in the Antarctic?"

"What? What are you saying? I know that noise!"

"That was the year," Hans went on, "of the last known personal battery anywhere on earth, as far as we knew. That last battery died out in sync with the totally unexpected global disaster. Who would have figured? Well, of course, any one of sound mind would appreciate the insane terror of such an event."

He opened a rear cabinet with another key, and there revealed the source of the beeping.

"But there *is* a battery!" Petrus exclaimed, jolted, gazing at the small, entangled patchwork device that was clearly transmitting off any number of stealth towers across the world. "We heard this signal at our station. My mother knew from where it was emanating. We had no way to respond. But this, here, now. Dijver 12, 8000 Brugge, right?"

"Correct. Although the last postal or delivery services disappeared many decades ago. But here you are, still standing."

"Hans, this is a total game changer. Who else knows about this?" Petrus pressed anxiously.

"Forget it," Hans declared emphatically. "A few governments at the height of their turmoil still maintained some battery packs with lithium-sulphur electrodes and millisecond discharge. Then there were the rare cases of those who might manage to get hold of one of these, hybrid anode silicon carbon, basically a nanotube battery. Burdensome names but it weighs less than ten pounds, and it's been powering an international alert signal every minute, more or less. You can count it."

"What's this?"

"The hand-crankable old fashioned solar-powered backup, and a fairly reliable spark-gap transmitter for damped, local intermittent transmissions," Hans added. "Which was a terrible idea."

"We had stuff like this at pes. I was never formally trained as a technician, but we all shared tasks near the end. But everything was dead. If it works, that's obviously fantastic news!"

"I believe you would have survived and found your way back to Bruges, in any case," Hans volunteered.

"No. I was just incredibly lucky. I told you pretty much the whole horrible saga. It was timing, down to the second."

"You were destined, I think."

"Nonsense," Petrus said, his mind working rapidly to assimilate these new prospects.

Both men paused. There was a sudden and unknown space of reasoning between them. Petrus was laser-focused upon the contraption.

"Look, at the time we had no clue that it was all going to hell," Hans started. "Only normal instincts. Your father loved you and your mother. That was it. I couldn't say, 'No,' obviously. I cobbled together this, junk mostly. Whatever I could get my hands on. The early Emergency Position Indicators they use on ships.

Because, in the beginning, SARSAT was still up there, nearly 530 miles high, orbiting every 102 minutes. He had such hopes, your father. I didn't. That was a profound difference between us. My training as an astrophysicist made the extinction of whole worlds and grand explosions the norm. Oh, there's also the 121.5 megahertz lower power signal. We didn't know anything, in the beginning. But what really worried me the most were locals. Western Europeans."

Petrus was stunned by all the implications of this. "We couldn't respond at the station."

"Of course you couldn't."

"A few guys who knew this stuff jerry-rigged what they called foxhole and crystal home-made radios. But there was simply no way to find a repeater on any number of other networks, not from the Antarctic. Our radio people tried the obvious frequencies, I guess."

"They would have," said Hans. "The International Red Cross at 47.42, all the way to the obvious 121.50 aeronautical emergency frequency" [3].

"I believe so." Petrus' heart was pounding.

"And I'm sure your colleagues down there first noticed it when every NOAA bulletin went dead and never resumed. Right?"

"Yes."

"Even 259.70, the space shuttle communication frequency. Silent. Same with the military at 243.00."

"That's right." And Petrus recounted: "They tried everything, month after month, year after year. I watched people growing old and grey as they groped hopelessly through the static."

Hans just shook his head. "Petrus, there are no tenable emergency frequencies at sea, in space, or on land, not in the midst of these tenacious and continuing solar storms."

"I understand," Petrus said, frustrated by the old news. "All the endless talk of frequencies. But here it is. It's working!"

"No, I don't think you really get it," Hans went on. "How many others, very clever people in their holdouts across Europe, Asia, everywhere are fiddling with the archaic junk? Ham operators, hundreds of radio clubs. Why hasn't there been communication, repeaters like the Eiffel Tower. Or our own Belfry? Therein lies the terrible rub, Petrus. The CME/EMP storms are not transient. If you are a radio it's the equivalent of a massive nuclear war. We have to assume it will be like this for hundreds, maybe thousands of years. We're now speaking of geological time."

References

1. "Scientists Just Proved That the Humanoid Lamb in the Ghent Altarpiece That Everyone Made Fun of Is Supposed to Look Like That," by Taylor Dafoe, Artnet News, July 30, 2020, https://news.artnet.com/art-world/ghent-altarpiece-lamb-1898463
2. The Getty Foundation, https://www.getty.edu/foundation/initiatives/past/panelpaintings/panel_paintings_ghent.html
3. "Distress and Urgency Procedures," https://www.faa.gov/air_traffic/publications/atpubs/aim_html/chap6_section_3.html

Chapter 34
The Emergency Beacon

"So, what do you suggest?" Petrus finally said, weary of the jumbled history he had now hit head on.

"We just turn it off." Hans knelt down to do so, before Petrus intervened.

"Please, not yet."

Hans was taken aback, his resolve broken through by utter annoyance. Then said, "Look, your father asked me to keep it transmitting for one reason only. Well, one of you has made it. All those emergency frequencies that failed you, failed us. I wasn't sure about it. I explained to your father the risks. I had and I still have no reason to doubt that someone within Bruges, even, could have a hidden radio at the 2200 meter distance, a 1.35.7 to 137.8 kHz frequency. Fortunately, Stefanus did not know Morse Code."

"I do."

"No."

"What's the risk?"

They stood like combatants on a cliff.

"The risk," Petrus continued quite firmly, "would be to turn it *off*." His youthful energy frightened Hans. "That would be" he thought about what it *would* be, "fatalistic, self-destructive. There is no sense in doing that. This is a powerful beacon and a radio and it obviously works."

"Yes. But you're not thinking clearly, Petrus. You're young." Hans was deeply panicked. There was too much to explain and he was no force against this youthful intruder who happened to be family. "You didn't live through it all," Hans declared.

"You don't have a clue what I went through, Hans."

"No. But I did watch my brother, your dad, perish from Merkel cell neuroendocrine carcinoma of the skin, and associated ills. It was not all that rapid. He suffered. Like everyone else. He just couldn't face the moment. The abdomen was distended, the neck and face swollen, he couldn't breathe, the body deteriorating rapidly."

"Stop."

M. C. Tobias, *The Maiden Voyage of Petrus van Stijn*,
https://doi.org/10.1007/978-3-030-97683-5_34

"No, listen. He had no dialysis. No ventilator. No power. It happened not fast enough, as with some nine billion others, ten billion, no one can know for certain. Famine was widespread, *famine*, Petrus, as never before visited upon our species, especially across Africa and much of Central America and Middle East one can assume. It hit us death by death. But then, when there was no more news, who knows. I believe there were militia, crusaders. And they're out there still. Gathering force. We just have no historians to write of that darkness. And no idea what the situation might be at this very moment in, I don't know, Hungary, China. We waited night after night for invaders to come. We had no idea what to expect, when we would die, what was happening, either in Washington or twenty miles down the road. Your father was brave. He held out until the pain was too great. I couldn't have done that. I refuse to suffer like he did."

Petrus mulled over all this. Not expecting such conflict, nor having earlier had to confront his own ignorance of what was real, of what had happened to a world he never knew.

Hans also took a measured beat. "Many people who could do so simply used their guns on themselves. It does take courage; more courage than despair, I imagine. Or walked into the ocean, that would be hard for me. But there was also the one-pill-for-all weights and metabolisms. You could get it if you had connections. I have three dozen such pills. The wonder pill. Two is probably an absolute given. It contains a chemical that accelerates the combined effects of nembutal and secobarbital. I've watched over several friends who took it. Your father took it, in the end. The brain turns the body off within seconds. I've also gazed upon tens-of-thousands of human corpses. I can't describe what that smells like. All the friends and strangers alike that I helped bury."

Petrus could not quite react to the enormity of what his uncle was describing. His experiences were smaller ones. Personal, but also interspecial at a level of amitie and sophrosyne he could not convey. He felt no compulsion to explain his deeply embedded sense of other life forms he had known, immersed his life in, particularly the penguins. But also lichens. Albatross. Seals. Petrus was at a complete loss in terms of describing his real-life biography to a man he scarcely knew and did not completely trust, not now

"And the alternative," Hans continued. "To be able to wander leisurely in silence through one of the greatest art museums in the world. To have no light interference at night, nor the absolute mayhem of tourist buses. I'm speaking of millions of people who used to visit our little town annually. And then there were the cars bumper-to-bumper, and garbage trucks and rising tensions worldwide day after day. Not anymore, as you see. No more babies in pain. No more horrors. And we don't have to watch or read any news."

"We have to discuss this." Petrus was animated now. It had sunk in.

"What's to discuss?" Hans stated with -Petrus noticed - a total elevation above the topic.

Petrus immediately knelt to the floor and, using Morse Code-.

"What are you doing? No, NO CQ, don't go there," Hans pleaded -.

to no effect as Petrus urgently tapped the requisite: . .   / ... / - / .... /. /. -. /. /. - /
-. / -.-- / -... / --- / -.. / -.-- / --- / ..- / - / - / .... /. /. -. /. ..--.. [Is there anybody out there?]

Hans was silent. Then, "Terrible mistake."

Chapter 35
The Science of Restraint

After some time of not speaking, Petrus thinking there might come a reply, only to hear continued silence, Hans said, "We simply are not able to ever return to the way it was."

"You could have turned the beacon off."

"I told you the reason. He only hoped your mother, the two of you, would be listening. I explained the terrible risks to him. But he was a stubborn man. He believed in art, and for some stupid ass reason felt no risk in drawing attention to ourselves. You've now walked the Iberian Peninsula, and through France and Belgium. You've seen the state of the world, and our little community here, such as it is."

"Which underscores my point. What risk?"

"Marauders. Psychos. The greedy. The violent."

"Nearly everyone I've met could not have been more friendly. Glad to see me. They're all doing just fine."

"You're being naïve."

"I've seen almost no children. Frankly, old men and women aren't prone to marauding," Petrus asserted.

"It's dangerous to invite strangers into what is essentially paradise. Our paradise, if history be any guide."

"So you're truly worried there might be barbarians emerging from the gates of hell?" Petrus asked.

"If you want to put it that way, yes. You do the math at the species level. You're evidently good at that."

"I don't follow."

"With the majority of survivors now mostly elderly, there will be biological pressures on any remaining youth. What kind of pressures, under the conditions I have witnessed in my life time, one can only imagine. But that translates into mutational opportunism."

M. C. Tobias, *The Maiden Voyage of Petrus van Stijn*,
https://doi.org/10.1007/978-3-030-97683-5_35

"That is a valid point, yes. But why go with the worst-case scenario?"

"Because I've seen enough to know. I'm practical. I like my coffee in the morning just so. Human history is nothing but an indefensible quest for order, some little remnant of daily routine, of unsurprising uniformity, against the gorgeous but deadly chaos that is nature."

"And you don't think there is something possible, something better?"

"Better? Better than nature?" Hans gazed upon this nearly Apollonian figure of naivety. *"Something possible?"* he reiterated. "What? Demographically speaking, no way. Medically? Again, no. The Sun has offered no clues and if it has, we are unable to read them here on earth. We're there. As far as we're ever going to get. That's my point. Listen to my words, kid: We're there. This is it. Why risk spoiling a Bruges that is now at peace with its few dozen odd human inhabitants. Good people. Some crazier than others. But we have a doctor, a celloist, several cottage gardeners, like the two who work part-time for you. There is Henk, who tinkers with old parts in a potting shed. Father Jakob Grimmerhausen, our nutty superstitious Priest for all the churches in the town. You'll meet him. He leads a congregation, playing a 4-stringed rebec. He's an old hippy activist, but also a bit of a silly trouble-maker, with a congregation of maybe five. Sophie you've met. People like that."

"Sophie was nice," Petrus said. "Anyone who feeds birds."

"She's wonderful. You wouldn't know it, with her wild hair and missing several teeth -we all are. Bad stomachs, most of us. Too many moles. The UV, of course. And no prospect of DNA repair, or nucleotide excision, that sort of thing. But Sophie was actually Belgium's Minister of Environment, long ago. Before that she was a historian of engineering. She could give an impromptu lecture on Volta or Maxwell. Or lecture any group of PhDs on Tesla coils or fuel cells. She knows the insides of hydropower and of an alternator. A vacuum tube and hydraulics, pressure waves and the voltage of sunlight."

"Seriously?"

"But all to no advisable end. Which doesn't seem to bother her in the least. Like all of us, she's resigned. She'd be the last one I'd want to know about the radio. As I said, she was a politician, and a good one. I'm not sure what she'd make of it. Her first instinct: broadcast. But then, insist on silence."

"That's an odd concept."

"Which thing?"

"Resignation," Petrus repeated. "Especially from a politician."

"Rather spiritual, I'd say." Hans could read his nephew. Something like skepticism or plain fundamental restlessness. Or maybe he was simply in a sustained shock before the new world. He'd never grown up with others his own age. He'd known only what Hans could imagine to be something akin to the surface of the Moon. Of course, he just didn't *really* know. But he was right about most things. Petrus was still trying to get his terrestrial sea legs. To achieve some order of familiarity with a green-lit biosphere. He knew nothing of war. Of true Hell. A dreamer in a bubble – a dangerous combination.

Hans felt compelled to remind him. "There is no satellite communication, Petrus. I repeat. You understand what that means. No electrical grid, anywhere that we know of. No infrastructure. And nobody cares. That's the thing. We've lived with this long enough to have forgotten the comforts and relish the simplicity, not that a castle, or my five-story mansion are not something to crow about. They who remain just want to be left alone, to be happy. Grow their gardens, be their eccentric selves, forget the old days with telephones and movies and nice cars and, admittedly a real frustration, ice cubes. And die quickly. Which is always a puzzle. What to do and when to do it. Natural remedies were always primed to keep ungainly people condemned to short life expectancies. By superseding those ancient rules to some extent, more than doubling our lifespans, only to be rudely cast back among the old rules, yeah, we royally got off course. It looks pretty from the outside. But when you get to be my age, yes, everything takes on a sterling simplicity because of all the hindsight, wanted, but mostly unwanted. And on any given day you measure the pain against the prospect of no pain perhaps tomorrow. Or a fine glass of wine. Or even a chocolate. You just don't know when to take the pill or pull the trigger."

"Are you sick, uncle?"

"My stomach. There's blood in my urine. But there has been for over a year. I mostly ignore it. Have to."

Petrus acknowledged silently. "Do you take something?"

"Of course. Almost everyone does. It helps with the pain, but something's wrong. I know it. Only a matter of time for me. Ultimately, we can't escape the Sun."

Petrus gestured towards the windows where the waning light revealed so much beauty in a perfectly framed steeple beyond a canal. Then he wondered aloud, "If the beacon has been going on for all these years, and no one has responded to it, you don't think perhaps you're being just a bit paranoid?"

"I'm tired of risk. Give an old man a reason not to fear yet something else."

"If there's so few people left on the planet," Petrus started calculating, "I really think it's a bad idea to turn off an emergency beacon. Maybe we, the people of Bruges, could give aid, comfort to others? Unselfishly."

"Aid?" Hans wondered with an addled voice of an old skeptic. Not a little amazed by his nephew's idealism.

But Petrus had already bent down and lifted the beacon. "I'll just hold on to it for a while."

"Turn it off. There's no emergency. We don't need it, we don't want help for Christ's sake. You're asking for trouble."

But Hans was no match for his zealous nephew, in every respect but logic and experience.

"Be stubborn, then. But you must conceal it. I insist." Hans found an old satchel Stefanus used to carry to work, in which to deposit the ten-pound instrument.

Chapter 36
Ethical Choices and Fine Wine

They wandered back from the museum to the castle. Petrus carried the satchel over his shoulder and upon reaching the large dining room, he set it back up in a corner. The beacon was still transmitting in a low, regular ping, as it had evidently been doing for so many decades.

Then Petrus and Hans sat down to a dinner. The meal was waiting for them on esoteric plates from various historic periods, placed impeccably upon the table and prepared earlier by Ms. Girard.

Petrus stared fully aware at the out of kilter craziness of this picture before him, shoving the heavy, solid, ancient oak chair under his butt closer to the ridiculously long table. Lavish excess in a world of corpses and skewed biology.

"The fresh greens were assembled by Rembertus and Dom, and you have Ms. Girard to thank for this lovely spread. She's quite good at it." Petrus already harbored doubts about all three of them. Something felt sickeningly wrong. A schizoid truth smack in his face.

"Where are they?"

"I imagine all three have retired for the evening."

"Where?"

"To their respective homes. No modest maisons du guardian, mind you, lest you start behaving like some aristocrat. They each inhabit, not own, occupy, historic large canal houses in town," Hans explained. "They had their pick of the litter."

"Ownership is no longer in fashion?" Petrus asked.

"No longer *relevant*, unless you are either without a gun and bullets, or with no miracle pill. Such would be the peripheries of our last tinge of an economy."

"Nobody else lives in this castle?" Petrus asked.

"Your father was admired. Anyway, I wouldn't live here. Too drafty. Too many rats."

With that, both men settled into the moments where food is passed and the old rituals of napkins, the pouring of glasses, and, in this case, rather struggling toasts are initiated.

© The Author(s), under exclusive license to Springer Nature Switzerland AG 2022
M. C. Tobias, *The Maiden Voyage of Petrus van Stijn*,
https://doi.org/10.1007/978-3-030-97683-5_36

"This wine is not local," he advised, filling Petrus' groot wijnglas a second and third time. "It was a favorite of your father's, a Chateau Lafite Rothschild 1992, a Bordeaux Red blend from the Pauillac area on the Medoc peninsula. He kept several cases from that year."

They feasted on a Vichyssoise with roasted cream of wheat and smoked sprouts, a pasta doused in coconut Bolognese, a dish of quinoa in cheeses cured with paprika and leeks. Lettuces with turmeric, red onion and avocado, and an aioli sauce of ginger and feta. Everything had gotten cold but it certainly didn't matter.

"I have never had a meal like this," Petrus declared. "My God. Who eats like this?"

"What did you grow up on down there? Marinated snowballs?"

"I remember the time just before the end of the days when we could count on trade with other research stations. Then it all dried up. The greenhouse and hydroponics saved us, and pickling jars. Otherwise, it would have been strictly Pemican, Muesli and Porridge, some cheese. Lots of hot chocolate. Mom was vegetarian and raised me as one."

The four poodles were impatiently waiting.

"What do they eat?" Petrus motioned.

"Your father was also a vegetarian. Your mom's influence, I imagine. And he always fed them from the table."

Petrus slid slabs of the backed quinoa cheeses onto four plates near his side and the dogs made haste of it. Petrus cleaned his hands on the fine linen napkins.

"You think they'd eat meat?"

"No. Nor are their would-be victims complaining."

"What's happened, Hans. Dodos, aurochs, priceless old master paintings on the walls, but no thieves, no armed conflicts, no economy, no muscle-man mentality. Yet here we are drinking great wine, in a furnished castle. Am I dreaming? There is no ice in the water, but armoires throughout the castle filled with rare objects. I guess I'm just not used to it."

"Correct. But that's what's happening. All of it. Your father was always an odd mix. Loved old porcelain tureens, Delft pottery, soft paste Sèvres, like that Vincennes baluster pot and the bleu celeste ice-pails. Don't ask me why. He was obsessed with stuff that could easily break. He loved those bird-painted flower vases. An art historian, after all. You don't seem to take after him, in that sense."

"I guess we'll see. I'm still trying to absorb it, uncle. If I can call you by that."

Hans grinned. "I like the sound of it."

Hans stared directly into Petrus' glistening azure eyes that seemed to be drifting, utterly bereft, out in space. "Listen to me. It's been one series of disasters. But, sitting here, you add it all up and, well, it's not so bad, right?"

Petrus said nothing.

Hans continued, "I'm so sorry. About so many things." He grasped for the right countering, the rejoinder of some higher logic. Then, "but there is so much we've been given, now. I will go to my grave knowing that somehow inconceivably this has all worked out in the most miraculous of ways."

"Leibniz. The best of all possible worlds."

"Yes, but I don't think we're deceiving ourselves in some dark Platonic cave. We actually have honest cause for optimism."

"As a species? Earlier, uncle, you suggested anything but."

"As individuals. It's way too late for the species."

Chapter 37
Something Not Human

Petrus took in all that Hans had elaborated upon with a fully unaccustomed swig of the robust and resilient vino in his rose quartz honeycomb molded Rhineland rummer of the sixteenth century. The very feel of it conveyed to him a sense that his father could not have been more disconnected from the planet in which an Antarctica existed. Petrus didn't care in the least although the contradictions were free flowing. Grand designs predicating genetic cul de sacs, along with happiness, in the same beaker.

"Forget the species?"

"Yes," Hans soberly asserted. "Of course, you have to think beyond the so-called schizophrenic logjam of the old days if your intention is to live in spite of that, which it should be and I'm confident, if you take good care of yourself, will be. That wine should help in the endeavor."

Petrus' heart was pounding. Not because he thought he was going to die any moment, as was the case during the four-day storm, the wreckage at pes, and collapse of the ice shelf, or even his now mostly forgotten journey thousands of miles from the last continent with the impending certainty, every day, every hour, that he was, indeed lost. His heart was in a flurry as was his semi-inebriated brain because the stark reality was too bizarre to fully absorb.

"You're innocent, Petrus. How many people have you counted since evacuating the Antarctic?"

"I'm not sure."

"Twenty, thirty? Fifty? One hundred?"

"At the most. But that doesn't mean that much."

"Paris. You walked through Paris. Paris, Petrus. Last time I was there, you hadn't been born. The streets were still Parisian. Teeming. The old Paris. Now, the UV Index is routinely over 40, month after month. That's good maybe for microbial mats of blue green algae and the like. You'd know. But the only time historically, I should say recently, that we ever heard of such radiation was during a volcanic eruption in Bolivia back in 2003. And only at the mountain's very high summit.

M. C. Tobias, *The Maiden Voyage of Petrus van Stijn*,
https://doi.org/10.1007/978-3-030-97683-5_37

You didn't see people enjoying the Champs-Élysées or happily wandering through countryside. That's because they're dead. The few that have managed to survive have stayed indoors, mostly; had the right genes *and* have been *extremely lucky*. Lottery winners, each of them. You're a biologist, figure it out. The shock to the species was massive. It hit one and all virtually overnight. The stressors changed everything. Those of us who have survived have obviously somehow begun to evolve resiliency, adaptation. I'm just assuming. I haven't read the old stuff on your so-called modern synthesis, or any of the last populist accounts of genome editing and so forth from thirty years ago or more. But I do know that all the rules were finally broken. Not just by Osna. She was part of a scientific culture. So was I. Were we wrong to be inquisitive? I don't think so. Did molecular biologists become arrogant? Absolutely. Were there consequences to believing one's species superior to all others? No question, yes. Nature has come back with a profound vengeance. I don't understand, frankly, why you and I and Sophie and the few others are still alive. I'll shut up. I enjoy my own rambling."

They paused, both reflecting on the day.

"May I?"

"Please." Petrus helped Hans to another glass of the Bordeaux blend. Hans knew there was plenty downstairs where it came from. He also had plentiful antacids for his stomach.

At length, the silence between them grew a bit overbearing.

"Look, kid, my nephew, who would have believed. You resemble your dad, which is both comforting and very freaky. I don't mean to sound religious, or misanthropic, I'm not. But I do believe in some of the themes of the Bible and we have, by whatever quirky, erratic means, achieved, well, I'm happy to admit it, a version of true arcadia, as I tried to explain. I say that having spent those years of my career when it was possible to do so staring at millions of stars and other planets. So here we find ourselves in a world with just a few *Homo sapiens* left whose collective weight can no longer exert horrors on our co-fellows, or that's my hope. I can't be certain and I really advise against toying with chance. You understand. The monkeys and spiders and songbirds and whales. Hell, that's your field. I'm still happy just looking at the stars through my little home-grown observatory, really more of a primitive monocle on the roof. But it doesn't matter whether you can turn static, or a metal touching a tree, into electricity. Thomas-François Dalibard playing with conductor rods, lightning and a leyden jar, just north of Paris in 1752. Best to use a small jar, if you've never tried it. From where humanity is at this stage, I think it quite unlikely that there will ever be a sufficient build-up in capacity, infrastructure, know-how, or the youthful contingent with the energy necessary to translate these last fragmented holdouts of people into a working civilization. It's just not possible, in my view. Nor would it be desirable. There's never going to be another satellite network. And, as you can see, I'm very relieved about that. We mustn't mess with this scenario. It's too good a set of cards we've been dealt."

"What do you want out of life, Hans?"

"The little time I have left? Exactly where I am. We made it. As I said, we're the lucky ones."

Hans felt himself starting to doze off. Though he had not mentioned it, he had actually survived the same cancer that had taken his brother. Hans had come close to taking the miracle pill many times but always gave himself one more day, a paradox upon paradoxes that led him from the labyrinth of death to the unexpected survival rate of roughly 20% to live another five years. Something like that. Of course, that was assuming he stayed forever indoors. A rule by which he never adhered. Nor did he waste time meditating on it all. There had been enough doom and gloom. He had grown quite attached to four puppies he knew would grow up playing in the forest and each would have a high susceptibility rate amid so much UV. Now, with his very real nephew before him, he was feeling a high degree of susceptibility himself, to a connection that could easily self-destruct. He didn't want to nurture such emotions. He was afraid, and too tired presently to think through the mine field of family. And with it, the inevitable loss that was coming. He still had some healthy tears left.

Petrus looked suddenly at the emergency beacon in the corner of the dining room. Picasso, the largest of the poodles, curly red hair as thick as those sheep in the Outer Hebrides, was pawing at it and barking as a transmission was coming through. It had been less than three hours since Petrus had sent out the SOS. But this was coming over on a frequency neither he nor Hans had ever heard. In fact, it was not Morse Code nor Single Letter Beacon. Neither Chinese nor Russian. This was a signal that did not originate from the standard use of numbers and English letters. No "E" for example. And the strident, annoying pitch was not a radio frequency between 3 kHz and 250,000 MHz. In fact, from what Hans could hear this was a deliberate distortion: an evolutionary design feature decidedly manipulated by someone very skilled at deception.

The dining room was thick with an addled and frightened perplexity. Neither men had a clue what to make of this roughly 18 second transmission. Hackers?

It's as if it isn't really human, Petrus thought.

Chapter 38
And That Night

Amid a flurry of most disturbing half-flashes and aerial collapses upon the lit still-ness of his vantage, supine, heavily breathing, he glimpsed a far-distant object approaching over water. Coursing through whitecaps, dismissing other gigantic obstacles, the behemoth was coming straight on. It was a ship, armored with heavy timbers and metallic girding. A structure like a tight-knit Ticonderoga-class cruiser, or some more primitive, but nonetheless menacing attack vessel, striking a due West heading without so much as a tack, not the slightest flinch before the ember-burdened ocean.

It was several hundred feet in length, most certainly, fast, unimpeachable. Now emerged its Hominidae occupants, one by one, hairy, immense, broader than tall, donning metallic armor, like Japanese samurai, one male in particular, navigating. His tools were indecipherable but clearly abetted by antennae, cables, wires, and an elaborate deception of connected machines attuned to the job at hand, of heading the carrier straight-on.

Others gathered, one brachiating down along the ship's rigging. There were enormous decks, the nineteenth century-like batteries broadside. The sails shone zinc and lead, and many coal-like smokes floated from snorting orifices across the agitated waters. The ship was the size of 5 HMS Victories, or a Yamato Class battle-ship seizing with seasoned tirade against any and all detraction. Its occupants were creatures from some future time baring down on the present. Eyes glaring through a darkness in the day, a fire in the night. Their shoulders and heads were enormous, legs stubbly, truncated by some evolutionary device to ensure hips like those of a large brown bear. Their stiffly moving jaws articulated and mouths grunted, and all was horrible like the last day on earth when man was no more, eclipsed by some terrible new transmogrified monster, seeping, the wounds, of wraths.

They were coming on with unrelenting speed, outpacing the sunset, mocking the sunrise.

© The Author(s), under exclusive license to Springer Nature Switzerland AG 2022   141
M. C. Tobias, *The Maiden Voyage of Petrus van Stijn*,
https://doi.org/10.1007/978-3-030-97683-5_38

There were females among them, obsessively heavy nipples exposed beneath arrays of sunstars bouncing off irregular jewels. Their heads were shrunken, faces albino and free, blue hair streaming in dreadlocks, bloodied legs no more than sinews made for scampering fast and covered in fungal fuzz.

Until the deliverance of this message could—Petrus—take it no longer, awaking in the steam of his inward terrors and a sudden memory of crashing into the glacial megadune, and the image of pes a mile away through lifting ice fog, on fire. The world having caved in all around him in the perdurable night. An entire scientific station plunging hundreds of feet.

His time-keeping watch had been ripped off during one of countless disasters in previous months. What time of night or day it might be. Or where he was.

Disoriented, Petrus arose from the luxurious royal bedchamber with its elaborate complex of cream smooth quadruple sheets atop and below an abundance of other bedding materials. A synthetic ermine-like fur that draped the four-cornered canopy above. And silly slippers to move across both stone and teakwood leading to a privy that was nothing like the toilet hole in the chateau he had visited en route in France. This was a modern bathroom with a name, Extra-Sized Quadruple-Low-Flush Odor-Free Recycler, manufactured by some firm in Antwerp once, specializing in *chambre d'ablutions*.

Into which Petrus now commenced to dry heave and heave, until the wretched contents, mingled in a vomitous explosion, doused in the Bordeaux which he had vastly overconsumed.

Chapter 39
That Day

Dulce Roso had to make the decision of her life, she explained to her neighbor, the very old poetess Isabella Coronado, who knew everything. She knew that Dulce's tummy was enlarged, with no effort to conceal it. She remembered the stories of her own mother, who had agonized over being agnostic, and having been *accidentally* impregnated. What to do?

Her two burros, Hiccups and Iris, accompanied Dulce to her neighbor's most pleasant farmstead that morning, while Margarita, the lynx, remained at ease and steadfast beside the entrance to Dulce's house, pleasing herself watching all the butterflies, occasionally flipping her enormous five toe pads, lost like bunchgrass in the softness of her grey fur, at them.

Hiccups and his little blind companion followed Dulce everywhere and dreamt her dreams, a late twenty-first century couple of Plateros. Hiccups was by now at least twice Dulce's age, stunted, slightly twisted about, with enormous floppy ears. Anyone who saw him called him *floppy*, but Hiccups never deigned to notice such stupidity. He loved people, lizards, big cats, elephants, pythons, hippos, and especially shady groves, all alike. But he proffered a special love for Iris, to whom he showed nothing but absolute fidelity. Burros are like that. A gratitude tantamount to absolute dependence upon Dulce, who had saved him years before from quicksand by a remarkable combination of a telephone pole, a rope, Hiccups' desire to live, and timing. Ever since that near death incident, Hiccups had what the local equine guru ascribed to "thumps" but later on, it was clear that these were true, human-like hiccups, not continually, of course, as that can be the end of a burro, either because of unrelieved dehydration or peculiar calcium deficiency. But whenever he got scared or even alarmed, clearly flashing back to the traumatic event in the slimy sand, hiccups erupted.

Dulce knew how to quell such bouts: with a few bananas. That and hugs. It was so simple. Hiccups had three possibilities each night: either sleep curled up next to Iris, Dulce, or beside Margarita. Once in a while, a sheep joined them. It always depended on gas. Margarita would not tolerate burro or sheep gasses, so it really

M. C. Tobias, *The Maiden Voyage of Petrus van Stijn*,
https://doi.org/10.1007/978-3-030-97683-5_39

came down to that day's food consumption, and Hiccups loved to eat every other flower, and, of course, whatever Dulce was eating. Iris followed the rules set down by Hiccups. Some flowers, like those of the rhubarb, and any quantity of legumes, would produce more burro gas, and should be avoided for all concerned. As well as black nightshade and groundcherry.

Now Hiccups could see annoyingly that he was not to be the morning's center of attention, as Dulce and Isabella had much to discuss. Of course, Hiccups knew what was going on. A burro his age grasps everything, both in minutest detail and largest panorama.

"It was that very nice stranger, obviously it was," Dulce reminded her.

"I know. He was handsome, no?"

"Of course. But it has been three months and now I either go to see Tejada or I deal with it."

"It is no *it*, my dear. Tejada is a good doctor. He'll know what to do."

"I know what must be done. *He* is gone. Forever. I was reckless."

"You were in love."

"Please."

"Are you? In love?"

"Isabella, we are adults."

"No. I am an old woman with a few lean books to my name no one ever read; a spindly crone who should have died during the crisis instead of so many better people. I suppose I can say honestly my very mediocre poetry kept me going."

"It is beautiful poetry. Publishers thought so."

"Stop it. The point is, you are *not* an old woman. And the crisis, so much suffering, has made you immensely powerful. You had strong parents, not strong enough. But, like myself, you learned that it was possible to survive. And so you have. Now is not the time for hopscotch or checkers. Now you must help create our future. I am the walking dead. You are not, girl."

"I feel like a fool, and, honestly, I am frightened. I can feel the baby moving now."

"That's normal. What is not normal is that you may be the only pregnant woman in all of Spain."

They both were only too aware of the fact that the stranger had been the first young man either had seen in many years, and Dulce was probably the only young woman for hundreds of miles around.

"So what do you think, really?"

They talked on this way for nearly an hour. Dulce cried. Hiccups snored in the hot midday Sun, flies buzzing over him, but knew the clouds were being blown in, and the feeling was one of his favorites. With those clouds arrived the normal late day storms that had been routinely gathering over *La Sierra de la Mosca de Cáceres* every afternoon.

"Where was he headed, did he say?"

Dulce remembered precisely. "Far away, Isabella. Belgium, close to the Netherlands. A village."

"Which village?"

Dulce could not recall. She looked over to Hiccups, who had awakened with a toss of his enormous head at the flies.

"What do *you* think I should do, Hiccups?"

"You don't think he'll come back?" Isabella prompted.

"No, he was preoccupied. On a mission of sorts."

"I know. I saw. Remember? You introduced me, for all of one moment. I would like to have spent just a little more time with him. But I could see, you were instantly infatuated."

"Stop it, Isabella."

Dulce, seated in her plain cotton twill brown walking skirt and blouse, pulled arms around her head, face downward into her knees, her body shivering, and spilled into a new level of unmuted tears. She really had no idea what to do.

Chapter 40
Miracles

At roughly 5:30 am Hans climbed the 366 steps of the Belfry to watch the sunrise over the canals of Bruges. He had been for many hours in the exhausting grip of a nightmare, but also had felt, upon waking in a sweat, the premonitory sense of temblors all around. But he could not distinguish between that unsettling intimation and his own nerves. He routinely climbed the narrow vertical chasm of palimpsests dating originally from the early thirteenth century (famously added to, burnt, in part, and rebuilt over time) for badly needed exercise. This time of the day there was no greater view in this world, he thought.

At stair 302, something astonishing happened: More than 27 tons of bells just above Hans began to thunder. The acoustical throes delivered ear-puncturing punches, each 2–3 s apart in terrifying swings of percussion delivered like continuous lightning that could surely have been heard for a 100 mi or more beyond the central Bruges Markt laid out far below. Hans plugged his ears and continued with gasping breath towards the summit wherein he saw for himself what was going on.

There was *no one* there, no automated carillon, no wire mesh protecting the instrument, no rope railing that might abet a few large tricksters.

His ears tightly sealed to the extent possible, the cosmologist looked to the sky and there witnessed a meteorite shower that lasted over a minute, something he'd never before noticed at dawn. He tried to connect dots, for which no line between could be conceived. There was no palpable earthquake. Certainly no huge gusts of wind, indeed the early hour was as still as the waters in the canals below. Nothing to propel these enormous cast iron monsters. The astronomical event, he thought, might not be a coincidence.

"Someone there?" he shouted out between the swings of the many pendulums flooding the world around him.

M. C. Tobias, *The Maiden Voyage of Petrus van Stijn*,
https://doi.org/10.1007/978-3-030-97683-5_40

Across Bruges at that very instant, or more or less, a large, determined man, Jakob Grimmerhausen, his face profusely sweating with some overheated epiphany, fell to his knees beneath Michelangelo's 1504 white marble sculpture of the "Virgin and Child" in the Church of Our Lady, whose tower remained the highest structure in the entire town. A lightning rod. He would swear that the Virgin was weeping.

Chapter 41
Town Crier

"Madonna of Bruges," 1504, by Michelangelo (1475–1564), In O.L.V.-kerk Museum (Museum of the Church of Our Lady), Bruges, Belgium © Saliko, Public Domain

© The Author(s), under exclusive license to Springer Nature Switzerland AG 2022   149
M. C. Tobias, *The Maiden Voyage of Petrus van Stijn*,
https://doi.org/10.1007/978-3-030-97683-5_41

When Hans arrived at the castle, Picasso, Sartre, and the van Eycks greeted him and led him around back to the greenhouse where Petrus was exploring the rows of brilliantly flowering herbs, fruits, and vegetables. He'd never seen white asparagus, dragonfruit, or Okinawan sweet potato.

"You heard the bells, I presume?"

"They woke me. I'm surprised the town still has a bell-ringer."

"It doesn't. I suggest you join me at the town square. There's a commotion you should see."

They left at once for the square, dominated by the Belltower, Cloth Hall, the Provincial Palace, the Cranenburg and Bouckhoute Houses. A marketplace dating to the tenth century that many poets have long described as the most beautiful such square in the world.

When they arrived, Grimmerhausen was standing beside the large monument to the Flemish patriots of 1302, a nationalist sculpture fashioned by Paul de Vigne in 1887, as hundreds of millions of tourists over the years had been instructed by guides. There were as many as 20 people there. This time, the vociferous clergyman was without his customary musical instrument. His run-on-voice was intoxicated with the mad-man's mouthpiece of some celestial muse.

"Most of the town seems to be here," Hans told Petrus.

"Welcome Hans," Grimmerhausen declared for all the public to hear. "You and your newly arrived nephew we've heard about. A young person, quite amazing to see, to join us for this day of miracles and God's injunctions, wouldn't you agree?"

Hans had heard about the weeping Virgin on his way across town to get Petrus after his paralyzing moments in the Belfry. He had raced to see the sculpture for himself: not a drop of moisture.

"And what would those injunctions be?" asked Hans.

"We must, those who are up to it, march upon the Brussels Palace of Justice and inform the Court of Cassation what has happened."

There was no enthusiasm among the gathered crowd for marching anywhere.

"And what has happened, Jakob?"

"The crisis has ended. Christ has risen. Alleluia!"

Grimmerhausen, the defrocked priest who managed to amuse everyone to death, saw the skepticism in Hans' face, a scientist who thus embodied the man of the cloth's most delicious adversary.

"We all heard the bells, many saw the shooting stars."

"Meteorite shower," Hans retorted. "Eta Aquarid."

"Shooting stars and a weeping Virgin, our very own Michelangelo's. And I suppose you, Hans pushed the 55,000 pounds of iron with a few fingers at dawn?"

"No, but a sudden shift in the earth's axis did," the cosmologist replied.

"There was no earthquake. Did anyone feel an earthquake?" the priest lordy-lorded above the meager din.

"I didn't say earthquake."

"Only God could have awakened us in so jubilant, so redemptive a manner. When was the last time those bells clanged? I'll tell you when: Exactly 50 pre-adj years ago today, when the genie got out of the bottle and sirens wailed across the world to warn us all. But today we are liberated. Yes, as if the Sun itself had finally been subdued, from Joshua 10:12-14: And Joshua commanded the Sun to stand still" –.

"Upon Gibeon," echoed Sophie. "Yes, we all know."

"And the Nazis destroyed. Here now, I declare the re-establishment of the true Republic of Ter-Piété!"

"If it helps him sleep at night," a voice from the gathered ones rang out.

Most in the crowd knew of all the wild legends about that place, and the tug-a-war of valuable church mascots, priceless gold talismans, *spiritual lightning rods*, etc. for many centuries between Bruges, Ghent and the Netherlandish town of Biervliet in Zeeland.

"Actually, Jakob the earth has been knocked off its axis for some time, well over a century. Where have *you* been? Today, it simply moved again."

"I don't believe that for a second," shouted the Priest.

"Knocked off his axis all his life," Sophie chided good-naturedly.

"Ridiculous science chatter to refute the will of God, I'd say," Grimmerhausen mockingly rebuked his long-time beerhall rivals.

"Tell him, Petrus."

"He's right. I am Petrus, son of Stefanus and Osna van Stijn."

"I'll be damned!" a friendly if normally withdrawn man named Solomon with baking powder lining his hands, voiced, coming up to greet him. "I knew your father well, didn't I, Hans."

"That's Solomon Lupus Basin," Hans said by way of introduction. "Finest baker in Flanders."

"Ninety-two years and cooking," he replied with a smile. "What did you make of the bells, young man?"

All gathered close to the first newcomer to their village in decades.

"I've come from Antarctica. The magnetic south pole has moved a thousand miles or more."

"Speak plainly," the crowd cried out.

"There's many geomagnetic poles, they are in flux, every day, every minute. My Uncle's right. But he's the expert. All I can tell you is, the earth has been shifting, and you all know that. Sometimes it's an earthquake, sometimes something else."

"There, you see!" the Priest re-asserted. "Whatever you want to call it, it is something else. It's called God's will, and I've seen every one of you at some point in your life, usually every Sunday -though not *every* Sunday" he reminded them "in one Church or other, praying to that same God."

"The God who killed all our children," Hugo Heems, once a butcher, shouted out.

"And my wife," cried the part-time dentist, Flip Verhelst.

"And elevated all the waters, and made it impossible to visit cousins in Portugal, or turn on a light," Olivier DeWinter, a half-blind one-time contractor carped.

I watched as the Virgin wept," Grimmerhausen insisted.

"I saw no weeping, not even a tear, and I was there just after you, as you know," Hans rallied.

"You can't tell me you didn't see what I saw?"

"You were imaging a miracle through your own sweat and tears," Hans tried calmly to rationalize with the evangelist, who was well loved for his own inordinate addiction to fine libations, morning, noon and late into every night. He usually held it well, if cumulatively. But he also comported himself with a center of gravity in his mouth, high above the steaming paunch, rhetorical and rehearsed. It was old news. But his Falstaff-like orientation to the world, and very picture of himself catapulted onto the faithful and faithless alike, both inviting and uninvited, won him over to everyone he collided with. What Jakob missed most, beside his family, and Hans knew this, were rugby matches on TV as viewed communally from Vlissinghe's, where one could scream at the winning or losing teams, pounding the table among peerage with a twisted, happy fist, and downing a beer in rapid sequence. This had long been the religion of life for a man like Grimmerhausen. The rest was merely the decorum that glued the interstices back together, a common trade that filled his appetite for at least some degree of status, a reputation built upon that social animalism which flowed through his veins.

While everyone laughed and threw in their own similes and epithets, Hans said quietly to Petrus, "I go easy on him. He lost all three of his children to the cancer. Then came to the Catholic church from the Anglican Communion. His mind works in allegories. But his life has, from what I've witnessed, been an utter tragedy."

"Anyway, it's over sixty miles to Brussels. And you must weigh, what twenty-five stones? I'd say it's way out of your league," Sophie gently mocked him, knowing that by her very insistence she might get him away from Bruges for a few weeks.

"It is a long walk, I've done it," said Petrus. "And I saw almost no one in the city, far fewer than in Paris."

"What my nephew means to say, Jakob, is that there'll be no judges in attendance at the court. You're testing the judgement of Don Quixote. Only riddles will come of it."

"They need to know the truth, Hans," Grimmerhausen went on. "Decent God-fearing Flemish should not their entire lives be locked in this purgatory of unknowing suspension. For centuries man was too busy to look inward and ask the first layer of pressing questions. That's what those of us who choose the canvass sack are tasked with. To stop and ask yourself. How many of you have ever stopped and asked yourselves? How many of you have ever even stopped?"

By this time the Sun was already a scorcher, UV levels soaring. People were drifting away. Not even a miracle or two could keep them out in the open.

Chapter 42
Genes and Molecules Awry

"What really made those bells ring?" Petrus asked, as he and Hans walked back together to the castle.

"I have no idea," Hans admitted.

"That axis business?"

"No. Of course not. I don't know."

"No, yes?" Petrus urged.

"It's a mystery. I can't deny it. But I *can* ignore it." Then, "At least you got a sense of the extent of town politics. Two big miracles, nothing. People just drift back home. It's nice, don't you think? The absolute height of late 22nd century civility."

After some time, passing from town out into the forested domains, Hans asked, "So how are you getting along? Settling in alright?"

"I saw a herd comprising both mammoths and mastodons this morning, heading north."

"You would have gone bonkers over the original sparks."

"What do you mean?"

"The judgement of the unjudged. A perfect comeuppance, the parliament of all our other noble fellows, quadrupeds, eight legged, hoofed, feathered, given to flying in the night or carrying about a pouch. That genie our nut case Priest referenced. Those days, the first genetic detonations, they were real. Though probably closer to sixty years ago. Miracles every one of them. No court of appeals to even question the ethics of it all. The demon had flown the coup. And it was, I have to say, wondrous."

"Who managed it, and how? I mean, what did people do, what did they think?"

"All the editorials, pointless against such earnest endeavors as those of Osna's. Followed by a basal metabolism of hubris permeating our species' blood vessels. Grand-sounding rewilding attempts with de-extinction gene editing. CRISPR and so forth."

M. C. Tobias, *The Maiden Voyage of Petrus van Stijn*,
https://doi.org/10.1007/978-3-030-97683-5_42

"My mom employed it therapeutically, though her lab was limited."

"So, you know. Dolly, the Pyrenean ibex, or bucardo, then the aurochs."

"I saw both on my journey here."

"The popular debate is now 150 years old. Heath hens, passenger pigeons, Tasmanian tigers, the Great Auk, anything that lived during the last 700,000 years or more. That was the so-called *harvestable* material. Then all I can tell you is that it followed upon an out-of-control trajectory, with sequencing that confused gene editing with motive; a manipulation of traits mirroring the superego of egos. By utter accident, idiosyncratic minority-rules won the day in terms of dominance. I like to think it was cosmologically inevitable, recessive genes favoring non-violent behavior."

"Was there some other masterplan?" Petrus asked.

"Surely there was. Though it clearly backfired, proving something quite incredible. But the cascade of catastrophes undermined any rigorous study of what was going on. Or that was the apparent consequence. The Übermensch faded away, leaving oases of innocence, the original world. Though I have my doubts. No more predators? I keep looking down at the red and the black ants. Yes, the saber tooth proved it could be done, I suppose. And then we had our own problems to contend with and lost interest. The fact it all went to hell, at least for our speciesl afterwards seems to have been forgotten from the overall equation. How you forget ten billion people dead from cancer, hunger, depression, is a curious thing indeed. Human nature is clearly capable of it. No choice in the matter, neither yes nor no. But the results have been fortuitous, and I say that not as someone detached. I lost everyone I cared about. Yet I remain grateful. Who can deny the incredible view of lions and lambs lounging in communion."

"I'll have to think about the implications," Petrus ruminated out loud. "I mean really, evolution, natural selection, without predation? Harmful mutations? I gather there is no data bank, anywhere, as to what is happening to the populations of other species. Carrying capacity, boom and bust. It's a wonderful dream, but I don't have the knowledge to even guess how evolution could switch so quickly, and cope with the marginal, the excess, over-population amongst starlings and termites, krill and phytoplankton, that sort of thing. It's just not in my frame of transfinite logic. But if it's really true, it's wonderful, obviously."

"Grimmerhausen's God came up with good and evil. We spent thousands of years experimenting with hostility and mass murders. So, I'm not at all adverse to seeing gentleness resurrected. Anyway, what's natural is no longer up for debate. Everything is relevant. It's all part of a game plan and we're obviously the few lucky ones to be able to see it all unfold."

"I've thought much about that. As biologists we knew of synaptic interference syndrome."

"What's that?"

"Inferences mostly from bird communication that were easily deflected by background noise, clearly human noise," Petrus began. "Now, given what we've sensed might well be the total depauperate human population, that noise is gone. Which will not only help other species communicate again, but it's likely to re-establish empathic parameters. If, as you and others have suggested, we're down to 25,000 or less, then there should be no more interference. We should be able to talk with the animals. To feel them, and they, us. That Doctor Dolittle Gene, or molecular cluster, whatever it is. For the first time in fifty-thousand years."

"That makes sense," Hans said.

"Protein engineering. That and saving half the earth. Those were each the latest rage when I was growing up," Petrus expatiated aloud. "So, it appears that those endeavors were helped along by the Sun, and human vulnerability."

"And one can't blame anybody. That's the tricky part."

"Interesting."

"Quantum ecodynamics was just taking hold when I was got my Ph.D. at Cambridge," Hans began. "I still don't know what they were thinking. My own dissertation committee. So impressed with that idea that physics was the ultimate rational explanation. That we were capable of miracles."

"Like what?"

"Of seeing and understanding the beginning. Huge telescopes, no more ethical committees. Just do it. Most people saw it coming, whatever you call *it*. Tipping points, irreversible destruction, and so on. But no one, not down deep, ever knew what to make of their very own presence on a finite earth. The universe is sustainable with all its billions of stars and planets in billions of galaxies. But that doesn't quite play out here on such a little planet. And no one ever figured out how this demographic colonialism must inevitably wreak the very havoc that it did. Yet, evolution saw to it. Here we are. Say 25,000 of us. That's a mighty fine number. Was it chance? Purpose? Quantum roulette?"

Hans shook his head, unknowing, thought back momentarily, and asked with a deep change of topic, "I take it, then, that you are settling in?"

"I have no idea," Petrus replied nervously.

Chapter 43
The Van Stijn Paradox

For the entire afternoon, Petrus, wearing the thick anti-UV sunglasses wandered into the forest behind the castle. It was the first time in his entire life that he could actually immerse himself within the solace of shade trees. Be mesmerized by the feel of bark. Take in massively pungent soil. Immerse his hand in water that wasn't near zero degrees. Sniff the odoriferous. Gaze upon proud, unconcerned ducklings preening themselves in a garden pond before their watchful parents, no cares in the world. Or a red wild-eyed squirrel, peeking half-asleep in its nest in the fashionable fork of an ash tree. Dreaming of something. Its little paws shivering.

Everything before him was a life-altering novelty. So many epiphanies, in fact, added up to something akin to halo-hallucinating. Burying beetles, a Sumatran "burning ember" scarab, as the great entomologist Alexander Barrett Klots once described what he took to be the most iridescent beauty on earth [1]. There were aphrodisiac (E)-methyl geranate signals among hundreds of other exudations, a fanfare of carcasses and eggs beneath the roots of chestnut trees. Profusions of orthopterid fauna, assassin bugs, fawns browsing side-by-side with African wild dogs in a forest of two-hundred-foot-tall Douglas fir. And then, the European starlings ignoring termites and cockroaches, instead focusing for mealtime upon mosses, ground orchids, and chrysanthemums. Despite the continent's name, there were no ants in Antarctica, but here, before him, Petrus watched as scores of differing Flemish species of ants maneuvered through the loamy soil, carrying pieces of mastodon feces to their nests, both in the ground and within trees. Several blue- and yellow- colored ants were exploring his hand, but there was no biting nor stinging. Indeed, it was only when a large feline, an enormous spotted jewel emerged from the thicket and stared at him that Petrus was abruptly jettisoned from his reverie.

The cat was, no doubting it, a true Homotherium, a massive golden saber-toothed. The two stood gazing at one another, Petrus seated on a log teeming with termites, the Smilodon resting on his haunches as if it had been trained to do so. Like any child, Petrus had grown up reading about such apex predators. The largest adults, at nearly 1000 pounds, were capable in a team of bringing down a 12,000-pound mastodon. But not this one.

At length, Petrus muttered, "Here kitty... Good girl."

The cat approached matter-of-factly and rubbed his face against Petrus' hand, its tongue tossing out a few of the ants, ignoring the four poodles with Petrus.

Okay, this is really happening, Petrus thought, then stroked the animal's velvety fur with its darkly mottled rosettes. The animal wanted to play, and when Petrus scratched its muscle-laden beige-hued chest it chuffed, 800 to 1200 pounds of purr. This went on for some time until an incoming flock of toucans, toucans in Western Europe Petrus observed, amazed, alighted on the fir trees all around the six foot-two human and the massive felid. The birds scared her and she ran off. Petrus watched the tiger's musculature rippling like a mirror of mirages, unprecedented strength that had given in to nothing less than kindness; breaking down every barrier to affection, rebutting and rewriting all those now obsolete theories asserting survival of the fittest.

One of the toucans abruptly landed on Petrus' right shoulder and started preening his long blond locks, quite pleased with itself, by all appearances, for having terrified a saber-toothed tiger.

Ordinarily, one might wonder, as Petrus did, how evolution could have been so rapidly manipulated as to unleash a command-and-control chain of events that affected behavior across both vertebrate and invertebrate communities. Hans had referred to the tameness gene in red foxes, an experiment long ago that had focused upon one gene in particular, the SorCS1 that was correlated, as early as 2018 with a level of receptivity to domestication. Osna had most certainly believed in it [2].

Could the biological world, whole ecosystems, exist without violence, or, to put it more neutrally, predation? Cell by cell? Atom by atom? Ridiculous but why not? Must merging always signal hostility? This was counter-intuitive, in fact. Hence, a biological equal sign that had actually been achieved, Petrus reasoned. It was far more problematic, if promising an idea than anything he'd ever contemplated as actually being manifested on earth. Could the price of far more than 99% of all human life destroyed have been the missing link in the annals of a peaceable kingdom given the obvious fossil evidence that purported to one killing field after another for billions of years, all before humans? At first, even second glance, it seemed to be an implausible theory.

Osna and her cohorts had passionately endeavored to technically enshrine a perception steeped in one particular ethical obsession, namely, that there should never have been a biological world in which suffering was ordained and programmed. The genomic programmers of the early twenty-first century had to have understood that ten billion largely predacious *H. sapiens* could never change their behavior in time to engender such a condition across earth. No form of government, let alone a constellation of indecisive mayhems, supported such cohesive decision mechanisms. There had to have been those who knew there would be vast carnage before simple cooperation, anthropocenic horrors prior to redemption. Surely, if this equation for the ideal had in fact been managed, did it encompass the geographical planet, or merely certain biomes? Bubbles capable of sustaining the experimentation?

While hundreds of square miles of breeding grounds of white-blooded crocodile icefish had been lost as more and more of Antarctica caved in, other brood basins, neurological networks, and ethological commons had flourished, certainly between Gibraltar and Belgium. There was no telling what might have happened to the earth's remaining neotropics, reef structures, wetlands, freshwater riparian systems, dry forests, and boreal forests. But Osna's team had indicated, at least in the beginning, that new co-symbiotic and mutualistic narratives were usurping previous ecosystems, in some cases at the landscape level.

That could only mean that the hotspots and megadiverse [3], transboundary ecoclines and smooth gradients, traditionally conceived in the manner of "biting off a yawn," as Carol Kendall once wrote [4] in a different context, were in fact harboring an augmentation of co-operative structures. Minting new fellowships. Mentoring the great gamble in all of evolution: the universalization of non-violence.

Could biochemistry, photosynthesis, any of the known several hundred *constants* in math and physics thrive under so logical a biological umbrella?

Yes, Petrus reasoned. Osna linked in from the fifth largest continent. This was a global phenomenon that she and colleagues had imagined. The first in hundreds of millions of years. Whoever had participated in this experiment had apparently succeeded in eschewing the parable of the Serengeti, of eat or be eaten.

Even the stodgiest prophets of the Bible had promulgated a stern message of Armageddon before the onset of a transfigured Christ and his minions. So many of the great classicists, including Lucretius and Milton had relied on this transformational journey that was at once paradoxical and dualistic. A universal Lazarus gene assuring every organism of a real-time Assumption, Transfiguration, moksha, Nirvana, a Zen-like free space. But only after enormous suffering, which would continue no matter what. Startling liberation leveraged on the back of diffuse despair. Anger might be mollified, but memories could not be parted as easily as the Red Sea.

But forgetting religion, Petrus could only marvel at the proof-positive messaging that lived and breathed in this forest behind his father's castle, beneath a relentless irradiation that did not differentiate, either among humans or any other species. What else was he to make of these miraculous goings-on? Biblical prophecy, or scientific and engineering breakthroughs for the few? The world had changed exponentially in ways that no one could write of, as there were so few readers. But Petrus' own mother had seized upon the brutal facts to accelerate her own conviction that some, at least, should survive. She was determined that there was a way to see that happen, opportunistically, efficiently and urgently.

He wandered back towards the house, where the four nervous poodles as well as the gardeners and housekeeper were all there waiting for him. The toucans may have intimidated the poor tiger. But the tiger absolutely terrified the poodles. This was not lost on Petrus.

"Goedemorgen," he said, seeing by their respective expressions that something was amiss, and tossing a nearby frisbee for the dogs to pursue. But they simply stood still, as if shocked by an invisible air.

"There's something beeping in the corner of the dining room," Ms. Girard declared. Her sense of urgency dominated all else.

"It's a radio beacon," ventured Rembertus.

"I remember them from early on in the crisis," Dom added.

"Am I to understand that you know Morse Code?" Rembertus uttered with a very mixed and anxious resolve.

"We are all quite surprised, Mr. van Stijn," Ms. Girard said with a sense of accusation in her voice. As if to assert, *You knew of this all along?*

But Petrus was more concerned with the dogs. They looked troubled, possibly ill. The Sun seemed brighter, even than usual.

References

1. A.B. and Elsie B. Klots. *Living Insects of the World*. Doubleday, Garden City, N.Y., 1959.
2. * 606283, SORTILIN-RELATED VPS10 DOMAIN-CONTAINING RECEPTOR 1; SORCS1, Alternative titles; symbols, SORCS RECEPTOR 1, HGNC Approved Gene Symbol: SORCS1, Cytogenetic location: 10q25.1, Genomic coordinates (GRCh38): 10:106,573,662-107,181,137 (from NCBI), https://omim.org/entry/606283
3. See "Biodiversity Hotspots Defined," https://www.cepf.net/our-work/biodiversity-hotspots/ hotspots-defined#:~:text=There%   20are%20currently%2036%20recognized,as%20 %22endemic%22%20species. See also, *Megadiversity: Earth's Biologically Wealthiest Nations*, by Russell A. Mittermeier, Cristina Goettsch Mittermeier, and Patricio Robles Gil, Foreword by Edward O. Wilson, CEMEX, Mexico, 2005.
4. See *The Gammage Cup*, by Carol Kendall, Harcourt, Brace And Company, New York, NY, 1959, p. 17.

Chapter 44
Leviathans

Because it was a non-GPS beacon, and in absence of anything like a polar orbiting satellite, Petrus knew, unlike the gardeners and housekeeper, that there was something else entirely going on with that radio concatenation in the corner of the dining room. But that it was a radio, with all kinds of newfangled attachments and odd gadgetry, and it *worked*, that was the news that would circulate throughout Bruges by nightfall. None of the three employees at the castle could keep their mouths shut. Nor had they any reason to think they should. As far as they were concerned, this was life or death, a radio that could, in their minds, remedy the longstanding "war of all against all," as Mr. Hobbes had once put it. Notwithstanding that for many years before, during and after the crisis, they had been among the lucky ones to survive, and had actually been gifted with lives. Yes, their ken had mostly perished. But here they were, standing firm.

Still, they wanted more. Not just telephones or movies, cruise holidays to the Bahamas, but more. Not more old masters, nor warm chocolate croissants or ginger snaps. Neither morphine nor Pepto Bismo, bandages or Brandy. They had such things from simply scavenging from empty flat to small and formerly large business, hundreds of them, whose previous occupants had perished. They wanted more. Neither traffic free roadways nor calm canals. Those were right before them. Not games of horseshoe or Krulbollen (rolle bolle). That was an everyday affair. No, they wanted something else entirely. Fine wines, a relatively normal Spring with lovely gentle rains upon vast fields of tulips and sunflowers. Not enough. Something other than five fascinating seasons and a few friends. Their litanies of hopes and dreams might never be satisfied, certainly not in their lifetimes. And, Petrus knew, they were not so foolish as to imagine that there was some higher power, somewhere in the world, that could by its authority change the levels of ultraviolet radiation from the Sun or bring back even some of their deceased children and grandchildren. What *did* they want?

They craved dental floss. Ice cubes. Air conditioning. Ms. Girard longed for a hair dryer. Dom dreamt of a "real" double-cheeseburger, as he put it. Bean patties

M. C. Tobias, *The Maiden Voyage of Petrus van Stijn*,
https://doi.org/10.1007/978-3-030-97683-5_44

and pickle relish were evidently tiresome. And he missed throngs of gaiety in the Markt. Just old-fashioned conviviality. He could not stop thinking about his late wife, his entire family, which had died of the cancer. He lamented so many things he could not remember them all, at age 84. The fact that he could sleep in nearly any bed in any mansion unquestioned, unconfronted, did not seem to matter. That all the money in the world was worthless to him, when a heating pad, or cold milk shake was unobtainable.

Rembertus had fine beers, chocolates, French fries, and Mayonnaise. But he, too, wanted more. At age 91 he never thought that he'd still be gardening and enjoying such an unheard-of miscellany of plants from around the world that had managed to find a foothold in the soils of Western Flanders. Confusing as it all was, he knew that his little universe rivaled Kew Gardens and the tropics of the Cameroon. Yet, he was not satisfied. He could lift a Van Gogh off the walls, and stroke its irises, or sleep on a mat beneath the Ghent Altarpiece anytime he pleased. Or do half-a-million other things which, just a century before would have been unthinkable. Despite all that, he, too, remained unclear in his mind what was best, or how to achieve it. These were not optimists, nor pessimists. They were bored, restless, certain that somewhere other people were happier, there were more effective medicines for arthritis, and the grass was greener.

That night, having moved the radio beacon to his bedroom, Petrus was awakened from a half-sleep by another transmission, this time similar to the last.

A subsequent nightmare repeated itself, as he watched invaders coming headlong towards the continent, if not Bruges herself. But unlike other prehistoric relics come to life with a casual and sweet-smelling air, these were horrible brutes, out to conquer those few remaining humans on the planet. As if they were riding herd over all evolution, certainly primate evolution. It could not have been a more realistic phantasm.

By morning, the Markt was swelling with rancid gossip, angry opinion and all the makings of a revolution that encompassed all twenty-six residents, as it turned out, of Bruges. Three additional men and one woman, a centenarian, were absent on account they were dying in their respective canal houses and could not leave their beds.

Happily, both Hans and Petrus thought, the mischief maker, Jakob Grimmerhausen, was not present for these goings-on, having abruptly deserted Bruges the previous afternoon to commence his solo Crusade towards the capital of Belgium, and there inform the highest court in the land that the Messiah was on his way.

But there was no question that a second commotion had erupted and this time it centered upon one thing: the existence of a radio, a battery, the possibility of transmitting information throughout the world. Somehow this fact brought back to these few remaining denizens of the great Renaissance village the entire history of civilizations; the undimmed power of the human spirit to get beyond itself, to live a fuller life, to banish boredom with hope. Even if no one knew what to hope for, or why.

They were, evidently, no longer interested in so much continuous physical beauty of the landscape of their richly historical town. Exhausted by the previous turmoil that had ruined their families, undermined their religious beliefs, destroyed their economic livelihoods, and wreaked havoc with any sense of continuity, not that a person wakes up on any given day and says, "I want continuity."

Yet, there was no doubting the malaise and apathy. Hans had certainly lived with it and he had secretly wondered whether there might not come a day when the emergency beacon at his brother's castle would just suddenly come alive. With ghosts calling from a new sort of grave.

But to think that such a relatively primitive device could inspire profound agonies of hope amongst these last survivors was a bathetic culmination of the most transparent century, ever. Decade after decade that had kept no secrets from scientists or laypersons. Anyone could see the ruination, feel the blistering crescendos of extinction, pollution, death, and decay. All had been irretrievably laid bare, even as the unrelenting uprooting of every ecosystem accelerated. As the cancers, famines, weather anomalies and melting icecaps spelled out the wreckage decades in advance for any idiot to clearly comprehend.

"What then, do you hope to achieve with a radio?" Hans, gravely pale with the idea, declared to all who had gathered in the square.

"Wouldn't it do to see how the rest of the world is faring?" Geert Gossens, a former interior designer, now in his eighties, wondered aloud.

"Doing? Ask my nephew. He just traversed half the planet."

All looked towards Petrus.

"Not to put you on the spot," Sophie added with her knowing grin. She and her buddy Hans had long been ideologically joined at the hip.

"That's fine," Petrus began. "And thank you for this forum. It's as close as I have ever gotten into what you might call *politics*."

"Not politics," someone in the crowd shouted with a mocking smirk, remembering times past. This newcomer had no clue, that was obvious to many in the crowd. He was lacking totally in even a fundamental ingenuity. They could not know a fraction of the sum of his parts.

"Just your assessment. Tell them what you've told me," Hans said.

"I traveled over 10,000 miles, past three continents, or 40% of the circumference of the planet. I tell you this: I saw the sum-total of fifty or so people, not including you good folks. But millions of others -albatross, petrels, marine mammals, massive murmurations of birds -that's when they flock beautifully. Hundreds of thousands of turtles, and snakes, of frogs and pollinators that had been cloned, I guess, and have rapidly evolved, some from prehistoric times right here in our backyard, as you all know."

"You probably haven't seen the cave lions yet," an old farmer declared. Then, with a berserker giggle, "Timid as a dormouse."

"I've breezed past the Cape of Good Hope. Slinked up the entire West Coast of Africa. I've slept at the Rock of Gibraltar, traversed much of Spain and France and Belgium and I can tell you with absolute assurance that you are living, here, in Bruges, in the best the world has to offer. In the meantime, the Antarctic has caved in. That's worth thinking about."

Of course, Petrus knew in his heart that he was speaking at odds. For he, too, wanted to believe, to hope for something, anything. For that's what hope is. A fickle monster of dissatisfaction. The enemy of ennui emanating from its own disturbed backyard. Or the very real gasps of desperation and suffering. Hope doesn't know what it's doing. Or where it's going. Or to what end. Never. Or how bad it could get. Only the abstract measure of the opposite of what is painful and useless. Yes, it could always get worse. Or, perhaps, it could get better. Hope, in fact, was always a dialectical dreamer, lost in so many ways, rarely found. Petrus understood his very own dilemma.

"What about Tokyo?" someone mumbled. "Or New York?" raised another.

"Partly under water, I would assume," Petrus opined.

"But you don't know for sure?"

"No, and what would it matter? Who cares. You're doing fine here. Better than fine." But he knew that there *was* a path.

"We don't even know what's going on in London, less than 180 miles away. A radio could change all that. Where did you come by it?"

"A family heirloom," Hans abruptly butted in, "and no one is stopping you from going to London if you are so keen on a journey. Boats at Ooostende for the picking."

"What do they say on the radio? Why don't you let us all see it and listen for ourselves. Surely it's not just static?" asked Bonaventura Adornes, a 90 + something humanist, as he thought of himself. There was not a day when he failed to write an annoying letter to the op ed. column, not that any such thing existed [1].

"There are no frequencies. No background noise whatsoever," Petrus exclaimed, defending a position that he and Hans, of course, knew to be a most peculiar untruth. "In other words, despite a functioning battery pack, there appears to be no other radio in the world."

"Why should we believe you?" Adornes went on.

"Why wouldn't you," Petrus defended himself, digging deeper into his lie.

"The boy has nothing to hide, Boni," as Hans called Bonaventura from long custom.

"Humor me, then. Let me hear it for myself. Why the big secret?"

Petrus looked to Hans, who shrugged his shoulders with a look of pure innocence.

"There's no secret. I'll be back then shortly," Petrus said. "Of course I will show you. And then hopefully this myth can be put to rest."

An hour later he had returned to the Markt holding part of the radio, the part that would acquit their doubts. And so he did, turning it on. All marveled at the first instance of a battery-powered device in over half-century.

Boni tried first, slowly exploring the full range of frequencies. It was dead and everybody heard that it was dead. And that was the end of it. Just like that. It readily shone a window on the evolutionary proof of proofs; the quintessential self-flattery inherent to daydreams and persistent hankering. There was no battery, no power, no God. Only this one earth, the static of cosmic background noise, as Hans explained, sending a clear enough signal.

"This is all there is," the cosmologist conveyed, fully concealing his displeasure at his role in a conspiracy to withhold what he believed a most terrifying prospect. Of course, he knew Petrus had left some critical machine parts back at the castle.

Reference

1. It should be noted that Bonaventura spent his afternoons in his private courtyard putting, as he perpetually put it, *the finishing touches* on some fantastic contraption, a hot air balloon that would allow him to circumnavigate the world and find out for himself what was really going on. He had explained to Hans that he had assembled the basket, fashioned sheets of rubber, cannisters of edible oil to fuel the thing, and was just months if not weeks away from launching his expedition.

Chapter 45
Dinner with Hans

Later that afternoon, Hans, Sophie, and Petrus shared some beers at the old brasserie, Tompouce, not far from the Belfry and certainly one of the best such quarters in the country. The ad hoc bartender, Levina Vasaeus, was once a pathologist. At the age of 23 she got first stage melanoma. It spread to the lymph nodes, her arms were ulcerated, and the lactate dehydrogenase levels in her blood indicated extreme metastasis. Yet, she was the one-in-a-million. She survived having come within minutes, repeatedly, of taking her own life. That was nearly 65 years ago. On this day, she could not have been in better spirits.

"Don't mind some of our fellow Brugians," she told Petrus. "They become enflamed just like the filariasis parasite after which our town has unfortunately been drafted to denote the rare Malaysian tropical disease."

"You're a doctor?"

"Long retired. Fortunately, I kept as many drugs as possible."

"And she shares them most generously," Sophie volunteered.

"We all work together," Levina applied. Then, "So, that radio really is static free?"

Hans cast his glance at his nephew who replied at once, "Yes, except for the chaotic background noises, which are normal."

"Which means?"

"That there is not a ham operator anywhere in Belgium or France or the Netherlands," Hans affirmed with no room for doubt.

"It means more than that, actually." Petrus went on to succinctly spell out the global implications. "There's just nobody out there who has even the most limited components of an infrastructure by which to respond." He was both hoping to quell, while simply betraying all his own question marks. Hans felt the rash conspiratorial duplicity to which Petrus had necessarily given in. He had come to recognize that politics might be deemed strictly local, silly and of no consequence. But it only appeared that way.

"We're really alone, then," Sophie concluded.

© The Author(s), under exclusive license to Springer Nature Switzerland AG 2022
M. C. Tobias, *The Maiden Voyage of Petrus van Stijn*,
https://doi.org/10.1007/978-3-030-97683-5_45

"No, not exactly," Hans aided his nephew in clarifying. "We've all known that there are people throughout the world, a few in Paris, probably in Madrid, in every city. But we have applied some theories of extrapolation to those meager forensics. Individual counts, like the one in Paris and the other in Brussels years ago. Our species is almost gone. We just have to accept that."

The silence was relieved by the beers.

"I've accepted it. We all have," said Levina. "Frankly, I don't know what we really expected. Life is okay. We've come to terms with no more microwave ovens, plastic surgery or ultrasound-guided lumbar epidurals."

Levina and Sophie both gazed upon the youth before them. Perhaps in their eyes they saw their own husbands, generations before; or, in the case of Sophie, a grown son who never made it. Perhaps they believed that this incarnation in a newcomer apotheosized the very nature of human existence, for a brief flash of time that obeyed no geographical or temporal rules, only nostalgic impulses.

That night over dinner at Hans' house, Petrus laid out the groundwork for a journey he had decided was urgent. He had to find Dulce. He was in love with her, this classical maiden of his dreams.

Hans had not expected this. "You'd bring her back here?"

"If she will."

"She will. Can you find her, though?"

"I remember most of the way, although it's true that I rarely turned round. Most of the vistas will be unfamiliar. But I remember which roads I took." And then, thinking of the future, "You'll look after the dogs?"

"Of course," Hans said, choking with sadness. "And none of your father's faithful staff are going anywhere. They're happy even if they do complain at times. They like getting heavy gold coins. An irrational pleasure, to be sure. We all complain. The town garbage dump is a disgrace. The last ophthalmologist died twenty years ago. My monocle needs repairs. All such things are impossible. There's little collective energy left to change or fix anything. Which is fine. You see the beauty of the town, even if, sometimes, beauty does nothing for one's cause. But what cause? There are no more. We've outlived all of them. Just watch out for the saber tooths at night."

Petrus described his miraculous encounter.

"Yes, but that was during the day," Hans advised.

"You're suggesting-".

"We have no record of encounters at night, which is when, if I'm not mistaken, most big cats have always hunted prey."

"The tameness gene. I saw ample evidence of it, in everything," Petrus decreed.

"Not at night. The night is different. Trust me on this. Okay, I'm no biologist. But I have common sense. Don't travel in the dark."

"It might take a few months, Hans. Are you going to be alright?" Petrus had seen and heard enough evidence already that Hans was indeed suffering. Postponing that inevitable and unthinkable moment of closure, when every eternity came calling.

"Amid the rabblement of human folly. Yes, I'll try to be here when you return. But no promises. I'm writing your dates down in my social calendar as we speak."

Chapter 46
The Radio

That night again Petrus had a repeated version of his nightmare. They were coming straight at him, knot by knot, navigating effortlessly. There was a brutal, single-mindedness about this terrifying vision and he found himself wrestling with it until his senses awoke in the aftermath of a terrible collision. He knew these nightmares were tied to the beacon.

He was breathing hard. He reached murkily for the bottle of rum he'd removed from a cabinet. It had to be 10 years old. A glazed twist glass, by candlelight, shone elegantly below the beautiful Petrus Christus portrait of a young girl, similar to the one in the nearby museum which had evidently been lent to the Groening Museum from the Gemäldegalerie in Berlin for an exhibition that was taking place when all the lights went out forever.

Signed and dated, 1471, like a female Pharoah, Hatshepsut or Nefertiti, she had most assuredly passed the mirror self-recognition test of history, with her slightly twisted lip that suggested uncertainty, modesty, possibly a slight hostility to the whole action of portraiture. Or foreknowledge of a dark European history?

Was she related to the famed portrait said to be a daughter of John Talbot second Earl of Shrewsbury? Stefanus had written a book about the Northern Renaissance and mentioned the circumstances of the royal wedding in Bruges in 1468 of Margaret of York to Charles the Bold, which was attended by a gaggle of famous hangers-on from the court of Edward IV, and all those upper echelons who hung out with the Duke of Burgundy. Through the dense craquelure Petrus remembered vividly the woman with whom he'd spent one night. The portrait was of a girl no more than 20 or 21. He had to find her for himself.

In the morning, just after dawn, Petrus prepared to set off with his old rucksack from pes, and just the absolute essentials: a bottle of water, chocolate bars, and some medicaments, including four euthanasia pills and a loaded handgun, which Hans insisted he carry at all times.

Then he chose to switch off the radio and stash it up in the attic amid suitcases, boxes of books and some family photographs, nick-nacks, and a few other fun sized

M. C. Tobias, *The Maiden Voyage of Petrus van Stijn*,
https://doi.org/10.1007/978-3-030-97683-5_46

things which he hadn't seen before. He sat down and examined them, vaguely recalling one of the spaceman metal toys. He'd had one as a baby in the Antarctic.

There were pictures of his parents together. He put one in his rucksack.

Out back he played briefly with the dogs, observing that they were all getting old, even during the few weeks that he had gotten to befriend them. He knew the Sun and the albedo effect were merciless towards any canine that low to the surface diffuse reflection.

He took a deep breath, left a message of thanks inside the entranceway for Ms. Girard, Rembertus and Dom, telling them he would be back as soon as possible, maybe within a few weeks. Not that he actually believed that.

Then he started towards town and the old freeway heading south in the direction of Lille and Paris.

Early that afternoon Ms. Girard told Rembertus how the radio was missing from the dining room but how she'd heard Petrus coming down from the attic. She'd been making a bed down a far corridor, unbeknownst to Petrus. Rembertus was determined, as was she, for whatever their desperate reasons, to find it and turn that radio on again, which they did. Petrus had never decoupled the emergency beacon. But their unskilled efforts would prove futile.

Chapter 47
The Search for Dulce

The first night out, he slept in an abandoned barn somewhere. Thinking to Hans' warnings. He was not exactly spooked, having spent a very different lifetime than that of his uncle, accustomed to the natural world in a way no cosmologist could be, he reckoned. Hans and Petrus' father were both urban minds, addicted to that aspect of human time. Petrus, conversely, felt happy to sleep on damp, old hay, and was pleased to receive a tribe of goats that came and slept in the barn with him after dark. They seemed less surprised by Petrus than worried about something out *there*.

But perhaps that was just Hans getting inside Petrus' head. Crawling, jumping and slithering things, spiders and the like only fascinated him, like the giant marine pycnogonida arthropods in Antarctica. He had no fears whatsoever. Not of anything. He was soon asleep.

In the morning, he scrounged throughout the long-evacuated kitchens of two nearby farmhouses and found edibles that would easily hold him for a few days. Then he set off on roads that circumvented any metropolitan clusters. He was a fast walker, capable of 25 miles without stopping.

After some days he suddenly deciphered far in the distance, the Eiffel Tower, making sure to circle round in as wide a radius as possible from the city. It wasn't that he was worried about it. He just had no appetite for strangers anymore, even the fine company of a Father Bruno.

One night, a pack of hyenas, at least he thought they were hyenas, came near. They began yelping frantically, playfully. He could see their shadows racing back and forth. They were restless, perhaps frightened, but not menacing. Agitation vied with curiosity. Petrus did what he intuitively believed pragmatic: he peed in large circles around his campfire. When the animals did seize up near, he counted 14 individuals, all eager to explore the fire, but his urine kept them at bay. They seemed mournful, almost tragic, as their silhouettes fleeted and flashed and ultimately the pack moved on. The stimulations of a rapidly changing biosphere. A pack of restive ghosts trapped in biological purgatory.

© The Author(s), under exclusive license to Springer Nature Switzerland AG 2022   171
M. C. Tobias, *The Maiden Voyage of Petrus van Stijn*,
https://doi.org/10.1007/978-3-030-97683-5_47

He found that sleeping in the abandoned barns was guaranteed to grant sleep. And there were always feral farm animals about to mollify any lingering doubts. If there were hens, and lambs and ducklings, there were no angry foxes or bobcats, or any other creature to break the dramatic wall in his head. That fence of reasoning had now easily formed around the comfortable notion that the world had been so genetically domesticated, from ant and mosquito to whole ecosystems, that there would never again be anything to worry about. That is, as long as you kept sunglasses on and wore a large brimmed hat -or, in his case, hoodie, during the day. It was to be that simple.

Nor was he ethically superstitious. Except for one consideration that had been bothering him. The red fox gene theory. It obviously did not take into account the fact that any human being is not just an individual, but a population. If those internal bacteria could turn so easily against the host, as they obviously had in the case of skin cancers throughout the species, and in any antibody mechanism, then the benevolence premise was shot to hell, at least in the case of this last standing primate. Ten billion of whom, more or less, had died prematurely. Since death could not be ascribed to a universal volunteerism, something had to be terribly wrong with the theory.

He put contemplation of this contradiction out of his head. But he could not do so completely. It was an impossible problem to solve, certainly while traveling by foot so many miles each day. When one is tired, or not feeling 100%, such concepts are irrelevant. The mind may reverse itself, or criss-cross. But it never starves to death.

One day he encountered an old farmer who offered Petrus a fine Madeira wine. The farmer said he'd not seen another human in over a year, by his count. As he reached northeastern Spain, there was plenty of cabbage and apples, and roasted cauliflower that could be enriched with oregano and mustard. A shepherd, in her 70s, mentioned something about Ash Wednesday. She offered him tobacco in exchange for his socks, an offer he had to decline. She complained of black bile and having not a single sheep or goat in her flock. Petrus easily recognized that she was dying. The moldy yellowing discoloration on her face, a heightened degree of gibberish. Her terrible body odor. It was only a matter of days or weeks, he reckoned, before she would be an uncountable statistic.

He gave her one of his miracle pills.

Later that same day he was approached by an Iberian wolf. He was sniffing flowers and came up to Petrus, wanting to play. He fed him some portions of lupine and wild chestnut, as well as pumpernickel doused in Mayonnaise.

"Iberian Wolf," Photo © M.C. Tobias

Later on, he was swarmed by huge flocks of Arctic and Savi's Warblers. Petrus floated in his mind with two mating bright rough-legged buzzards that swung balletic points throughout the high thermals. He observed Pallid swifts behaving strangely, a Lesser Spotted Eagle that came to his campsite for handouts, Horned larks and Iberian Chiffchaffs walking on a highway. At a fabled marsh, there was a plaque rooted in the rock that told of a German invasion sometime long ago. Probably a million or more *Quelia quelia*, the sub-Saharan red-billed weaver, were taking bubble baths around a waterfall. Were they migrating north? South? East? He kept comparing this cornucopia to everything he'd known previously and at moments, he had to sit down, almost nauseated by so much life.

Nothing could be reconciled, neither the short nor long history. The largely exoskeletal remains of what had been a human-dominated planet had, within a generation or two, been subsumed and overwhelmed by the wild creepers of a far more pertinacious and plausible reality. But it was surprisingly disconcerting, to Petrus.

He missed neither the stained-glass empyrean of Paris, the halls of genius at the Groeninge museum or the sated, everlasting castle. Nor had he a single remaining sentiment by now for the swirling, if sometime intimate pariahs that had been pes, and all the futile pursuits in the name of science which emanated from it. In fact, seated at the edge of the marsh somewhere just a day or two from the small mountain range he'd come to find, in transit between dreams of heaven and earth, he looked back upon his entire life as a comfortable wasteland punctuated by the love of plants and animals, nothing more. A Kingdom unto his own that had been informed by his mother, abetted by a few adult companions, and laid out between distant nunataks and brutally cold weather, no more. A life with no livelihood or reason. Nothing to do, day in day out. Just to survive seemed clearly to mark the nature of existence, with its anonymous lucky and unlucky souls. He was one of reputed twenty or so offspring of the Antarctic, a group of strangers, none of whom he'd ever met, who would quietly trace their silent ancestry to a white-out the size of a continent that meant nothing.

There was only one thing left to accomplish, something other than nothingness.

Chapter 48
Strange Memories

In a barn within a well-ordered but abandoned shrubland once known as Campillo de Deleitosa, he met his first marten and genet, two of the sweetest little creatures known to creaturehood. He knew nothing about them. They both cuddled up with him for the night. In the morning, when he departed, they both went in opposite directions, having eaten some crackers he hand-fed them. In the annals of biology such a little gesture would have no meaning. But to Petrus, this was the new world. He knew they once had been predators. How their morphology and behavior had been so rapidly altered, he could not divine. What were they,other than themselves? Perfectly happy, courteous, friendship seeking. Nor had he an opinion that was either weak or strong regarding such changes. It just was. But across the airwaves of his heart he was grateful to see no more harm, killing, ill-intentions. Only natural mortality, which must now mean solely disease, mortal injury, or suicide.

This then, he assumed, was the natural and good end of nature. Eudaimonia.

For many days he wandered along barren roads through the Sierra surrounding the ancient roman town of Cáceres, searching for the broken-down church in whose ruins he had first laid eyes on her. There was not a human to be found.

He arrived at a small series of caves one afternoon and upon entering at once noticed something glorious, a handprint painted red. It appeared supremely old and was rapidly decaying beneath a surface of acrid pink salt crystals. There had been plenty of vandals. Petrus had no way to know that it was over 65,000 years old, and once thought to be the oldest cave painting known to humanity, the Maltravieso. There were many dozens of such handprints, or former prints, as if an entire class-room of eager, innocent Paleolithic children had been tasked with stenciling their palms and fingers onto the cave walls. There was clear excitement among them. Astonished brevity before the gigantic unknowns. In a far corner of rock wall, on the moist turf, he knelt before two human skeletons, the bones of their hands clutching to one another. These were modern people.

Atop the cave were abandoned twenty-first century apartment buildings, a juxta-position that signaled to the explorer a new sense of time: how long would it take for

M. C. Tobias, *The Maiden Voyage of Petrus van Stijn*,
https://doi.org/10.1007/978-3-030-97683-5_48

those buildings to vanish, for sand and wind and flood water and minerals to erode their ugliness out of existence? It was unlikely, Petrus thought, that they would survive like the many handprints had, purposed and sacred within an infinite darkness that had once been of World Heritage importance.

Throughout the bleak hillsides he wandered. There were other caves, other petroglyphs. He knew from his readings growing up of the significance of cave paintings. But presently he came upon a scene that left him entirely bewitched.

Holding a torchlight which he fashioned from reeds, he crawled into cave penetralia that led him several hundred meters into the blackness. Suddenly a huge amphitheater had opened up. Petrus stood and shone his torch upon the walls as his predecessors would have done tens-of-thousands of years earlier. Portrayed across the granitic panorama before him were the most amazing creatures, not dissimilar from some of the hybrids he had seen thus far on his journey. Had all this been predicted many millennia ago?

Painted beneath the animal riot were dozens of human hands in various forms of prayer, mudras. The whole scene described an interspecies veneration of obvious import. Above the deftly painted ensemble were painted steps, ladders, rungs leading to another world. A new life. Hope was not just something born of the terrible twenty-first century.

Chapter 49
New World

Two days later, Petrus came upon a recognizable hamlet.

Dulce did not appear, but her neighbor did, the old one, Isabella Coronado. Her smile said it all.

"You, vagabond," she called out. "Tired of being a nomad? You might not even recognize her."

An hour later Dulce appeared, walking home with two jugs of water. With her was Hiccups and Iris.

"I thought you were gone forever," she said, her voice starting to break.

"You've put on weight," he grinned.

That night in her little farmhouse he learned that within 4 months he'd be a father. The question before them was whether to stay in the humble two room cottage; or attempt the return to Bruges. Dulce insisted she could make it. She had an excellent pair of boots, which offered much more solidity than the stupid worn through tennis shoes Petrus had been given when rescued near Gibraltar. But he stubbornly fancied them.

Dulce also felt the tug in her throat at the thought of leaving her neighbors, particularly the old poetess. They had watched out for one another for many years, ever since the death of both their entire families, prompting Isabella to seek out this little hamlet south of the city, where Dulce had actually been born, and watched everyone, every single member of the village, perish.

Now, she was uncertain about Hiccups, Iris and Margarita, as well as the sheep, until the next morning, Isabella assured her that she could easily look after all of them, as well as herself. Anyway, they had never had any problems looking after themselves. Just as long as human doors never shut out the animals from warm, cozy interiors at night.

"And the baby?"

"You're young, you're strong. So is the baby," Isabella explained to her.

M. C. Tobias, *The Maiden Voyage of Petrus van Stijn*, https://doi.org/10.1007/978-3-030-97683-5_49

So it was within a few days Dulce had said her goodbyes, kissed and hugged the burros and the lynx a thousand times, then some, and that was that. A large Entrefino sheep, part Churro, groaned. He skipped along with Dulce and Petrus for a hundred yards or so, then gave up and returned back to the others.

From a calm distance, sniffing flowers, watching the goings on, was a wolf, his tail wagging. Petrus knew it was the same one that had been following him for days.

Amid her tears, Petrus and Dulce began a new kind of voyage.

Chapter 50
Coda

"Barn in rural Western Europe," Photo © M.C. Tobias

M. C. Tobias, *The Maiden Voyage of Petrus van Stijn*,
https://doi.org/10.1007/978-3-030-97683-5_50

They were 4 weeks to the day, just 48 h or so from Bruges. Late one afternoon, Petrus and Dulce saw a distant barn, part garage, with metal cladding. It was oddly surrounded by barb wire. Petrus had not previously paid attention to such fencing in all his journeying.

Once inside, it was clear to them both that the straw flooring was infested with fleas, cockroaches, and tarantulas.

At once they decided they would go back outside to sleep in the meadow. However, the Sun which had earlier thrown off an unusually beautiful mellow light, was now obfuscated by fast moving lightning green skies. Then dark hail began to pour, the air heavy with electrical savvy. They had no choice but to move back indoors, knowing it would be a sleepless night.

But as they tried to sort out things inside the unwelcoming interior, watched over by a large European owl, Petrus stepped on something concealed in the hay. It snapped with a strident and hideous sound. The pain shattered his body.

His foot had gone directly into a hidden, spring-loaded, seriously injurious open trap that had been laid for an animal, a wolf, or even a bear. Or someone's fear of *any* big creatures in the dark.

The metal fangs had bitten severely into his right leg and blood was streaming down. He could not tell if bone had been damaged, but presently he lay there with the pounding hail and ferocious ball lightening, not sure what to do. He knew he was in trouble.

Dulce raked the remaining hay before moving one step, and the metal tools were everywhere, sharp, excessive, the possessions of some pervasive malevolence occupying the degenerate space. But there they were beholden to the dark, stormy night, trying to stop Petrus' bleeding. He took much of the morphine from his rucksack and kept consciousness all through the night.

Why would someone have laid a cruel trap? he wondered for hours. The fact of it began to topple a vast scaffolding of paradise-like fantasies he had been harboring about all of life, of a changed Nature. But in this, this fact was tantamount to total biological treason. He tried to shake off the logical implications, the philosophical betrayal, as Dulce, using torch light, was able to find gloves and tools that by adamant exertion enabled her to pry lose the vicious metal claws.

But what made no sense to either Petrus or Dulce was the intention behind the trap. What possible rationale in a world supposedly long beyond such imagined menaces, such deliberate evil? This would haunt them both, but particularly Petrus who continued to have his terrible nightmare in one form or other.

Limping, barely able to manage, Petrus and Osna spent three more days on the road. He took whatever drugs he had to stave off the apparent gangrene. One early afternoon, they arrived in Bruges and proceeded through the empty Markt, which he labored to describe to her as best he could, and then on to the castle.

In the nearly 10 weeks that he had been gone, something had happened. Ms. Girard, Rembertus and Dom were nowhere to be found. Nor was there any evidence that they had been there for many weeks. Moreover, Picasso, Sartre, and the van Eycks were not to be seen.

Beside the greenhouse, adjoining the grave of his father was a freshly placed stone above a slight and narrow row of upraised turf. There were flowers wilted there, as well as a little writing book. Petrus knelt down and opened it: The book, inside a small plastic ziploc bag, was blank, except for the handwritten words, "Good-bye, Petrus. And God Bless." It was signed by Hans.

"Did you know him well?" Dulce, who was demonstrably larger in the tummy, asked.

"Yes and no," Petrus said.

"Was he a religious man?"

"I don't know."

They both went to the enormous master bedroom. Dulce held her breath, not quite comprehending so lavish an estate. Wondering if this was all simply a strange and wonderful mistake that had to end. How could any of it be real? Tangible? Hers to occupy?

She felt her stomach and tried to imagine who she was carrying. And what a baby could possibly do in this world.

Petrus scavenged throughout the castle to find all the medical supplies needed to help expedite the recovery of his right leg, which had colored, and begun to swell as he had feared in the terribly high temperatures, for the Autumn. They rather quickly consumed a bottle of a very fine wine from Stefanus' cellar.

"You probably shouldn't be drinking alcohol," Petrus murmured, beneath his heavy veil of pain.

"It's okay," she teared, lightly stroking his face. She could easily feel that his forehead was too warm; this man who had been born upon the ice.

That night they both slept, more or less, soundly.

"As if Petrus was dreaming everything …" Photo © M.C. Tobias

Appendix: Brief Overview

Petrus van Stijn's world is besieged by two prime engines of destruction: massive geomagnetic storms caused by unprecedented solar storms, protracted coronal mass ejections (CME), and climate change wreaking unprecedented, but predictable collapse of the Antarctic ice shelves.

Petrus also has other problems to contend with, like surviving on a floating archipelago of ice, and then walking 1350 miles through a post-Apocalyptic world.

At the same time, Petrus will discover something of a true social and biological paradise. Herein lies the awesome paradox of a world where one species, ours, is facing extinction, while most others it appears—many genetically re-engineered—are enjoying a biodiversity renaissance.

Antarctic Crisis

The melting of Antarctica, and the collapse of its countless, enormous ice shelves has been under intensive scrutiny for at least four decades.[1] New data on the rate of melting ice turns out to be much worse than anyone had calculated.[2] Future projections for the last continent are anything but welcoming. Climate simulations examining "dynamic-thermodynamic ice sheet" models have shown that a current dearth of information pertaining to "meltwater input" and numerous unknowns with respect to "cooler surface air and surface ocean temperatures" has made extrapolations in the future particularly difficult.[3] Many believe the worst is yet to come. Others see that worst-case scenario as already happening, both in the Antarctic and in Greenland; with Antarctic emperor penguins projected to be extinct by 2100.[4] What looks to be clear is that melting ice from Greenland could eventually meet melting ice coming from the Antarctic, as occurs in the novel, where Petrus is rescued in the Atlantic waters off Gibraltar.

© The Editor(s) (if applicable) and The Author(s), under exclusive license to
Springer Nature Switzerland AG 2022
M. C. Tobias, *The Maiden Voyage of Petrus van Stijn*,
https://doi.org/10.1007/978-3-030-97683-5

Osna van Stijn's research, and that of her son, Petrus, involves a combination of long-standing interest from multiple scientific disciplines. Those areas of investigation include bioinvasives throughout the polar trophic systems, where dozens of Antarctic endemics are threatened.[5] In addition, Osna's work involves gene editing, and directed evolutionary prospects for resiliency and adaptation. She has a (fictional) lab at the Belgian Princess Elisabeth Station, the first emissions-free scientific station in Antarctica, its construction completed in 2009.[6]

Osna's work in the field of directed evolution[7] enjoyed something of an international kick-start when three scientists won the Nobel Prize for Chemistry for their work in this area in 2018.[8]

De-Extinctions and Re-Wilding

For years science and conservation biology have looked at bringing back extinct species[9] and quickly gaining traction not only in the realm of protected areas and contiguous ecological corridors, but in actually re-wilding marginalized areas.[10]

Many geneticists now argue that it is not a question of *whether* humans will be able to clone extinct animals, but *when*. And there is a significant debate in terms of the genetic drift (gene variant frequencies) and mutational opportunism in relation to (original) holotypes. Not to mention future habitat suitabilities for extinct species recoverability.[11]

Where the directed evolution discussions become incredibly speculative in the story has to do with the issue of a non-violence gene (the Doctor Dolittle Gene). Ethically, one can only imagine the changing rubrics biospherically necessary to pave the way for a world of herbivores, in absence of any predatory threats. But that is exactly the situation that appears to have emerged throughout the seventy-plus years of back-story in the novel. The author never gives away absolute specific years, only an approximate time frame that is further complicated by a newly asserted ecological doomsday clock. Calendar dates are either pre-adjustment, or adjustment, pre-adj and adj. There is an approximately 27-year difference between them. Every carnivorous instinct has been staunched, hostility gone; any form of predation beyond vegetarian consumption, not happening. The saber-toothed tiger, fire ants, insectivorous avifauna, and people have all been somehow genetically, if not molecularly modified. We never presume to describe the processes by which it happened, only that it did, miraculously, or devilishly. And there are some significant dangling modifiers that will necessarily lead us, in the end, to believe that not everything is as perfect as it might seem.

One of the most intriguing recent signs that behavioral phenotypes can be altered through selective breeding involving the identification of "genomic regions associated with the response to selection for behaviour" has involved *Vulpes vulpes*, the red fox, and the so-called tameness gene. It has been provisionally achieved in the highlighting of SorCS1, "which suggests a role for synaptic plasticity in fox domestication." And the researchers add that the fox provides "a powerful model for the genetic analysis of affiliative and aggressive behaviours that can benefit genetic studies."[12]

This resolution of ethical antipodes incredibly achieved during the worst die-off in human history (in this fictional account) with human breeding pairs reduced to the point of no genetic return, provides the fodder and baseline by which to explore and intimate some of the most pressing issues of our time. As individuals and the species, once at odds, now seem to have found philosophical and moral closure in an Arcadia that *Homo sapiens* will enjoy for only a few more generations, the rest of biodiversity at least appears to be thriving, in however a human-modified manner. But we're not sure of that. There are some ill-boding traces that all is not right in Belgium.

There is a relic, philosophical archaeology at play here; compounded by the fact that Petrus and Dulce are going to have a child, or children. Without many others like them, however, the human genetic cul-de-sac is all but guaranteed, or so we may deduce from the limited information domain posited in the novel.

Petrus' Uncle, Hans van Stijn, a cosmologist, knows this. He is a Renaissance man in what was the capital of the Renaissance in northern Europe, Bruges. But, in spite of his myriad harrowing experiences and penetrating insights into a thoroughly dark past, Hans's spirit will not be dimmed; he is an optimist. Just not about the future of humanity. His late brother, Stefanus, the Flemish art historian and father of Petrus, was also an optimist. He infinitely preferred the beautiful art history, flowers, and community of Bruges to the (as he perceived it) utter brutality of the Antarctic. This philosophical and sensory dialectic has its own psychology that we explore in the story. Humans contain every geography, and we all somehow attempt to express it, in our phylogenetic tree. The van Stijn family is both trapped and liberated by these contending ideologies and realities.

Hans and Petrus each suffer nightmares. They imagine some other "advanced" hominin on its way to overtake our species, the way we, in turn, outbid the Neanderthal for survival.

In a world that by all appearances is about to see the last generation or so of humans, what are the ethical standards that rule? Is there any singular oversight, or command control? Can any form of economy or governance remain? If not, do people still have hope of any kind of collective bargaining against the abyss? Or is their ignorance and absence of a crystal ball, their countless traumas of family loss and other tragedies, more than enough to have sobered them? Tamed their fears and aggressions? Are they docile, doomed, or fantastically free?

CME—Coronal Mass Ejections

The solar storms associated with sunspots visible from earth are nothing new. Sunspots were first documented in 1609. And by 1848 an astronomer in Zurich first started an index of daily sunspot counts. Today, the World Data Center for the Sunspot Index is located at the Royal Observatory of Belgium, where our character, the cosmologist Hans van Stijn used to do his work.[13] These spots have been occurring for billions of years. What is new is the enormity of the solar flares characterized in the novel, both the duration and intensity of such storms. Usually, at least those thus far documented, are temporary.[14] When they occur, usually in cycles that are predictable, the earth's magnetosphere and ionosphere are only briefly disturbed by events that include geomagnetically induced current that can affect everything from the planet's electrical transmission grids to transformers.[15] Most people have never personally experienced such storms, or have any reason, to date, to fear them.

Usually, such disturbances occur ephemerally, in durations from seconds to a few hours.[16] One disturbing indication that the vulnerability of power grids to such storms could go on for much longer periods occurred on March 13, 1989 when the Hydro-Québec power grid was fried almost instantly, causing a cascade effect whereby mitigation took 9 h, during which over six million people, more than 25% of Canada's total population, was without power.[17] Seven relays in the power system had been tripped. All GPS went out for approximately 10 min. There was a complete "information blackout."

Venezuela had a nation-wide 5-day blackout in March 2019. Power back-up failed as well. Hospitals, and every other power-dependent structure, shut down.[18] In 1921 there was a far worse CME which, had it happened today, would have taken out power to one/third of the U.S. population. And in early September 1859, two British astronomers measured a CME that has come to be known as the Carrington Event. It was the largest storm of its kind ever witnessed (although quite recently others have just missed earth by days). The International Astronomical Union (IAU) based in Paris designated it as SOL1859-09-01 because it began on September 1, 1859. The immense storm took less than 18 hours to reach the earth. Its auroras were seen throughout the world, and telegraph lines were burnt to destruction. The solar particles charging through telegraph lines, "shocked operators and lit telegraph paper on fire."[19]

Flare power is measured typically from A to B, C, M, and X (the worst).[20] The letters are like the Richter Scaling, each letter ten times more powerful than the previous one. No known Y or Z scale has ever been experienced on earth, but is so in this novel. The CMEs resulting in massive solar flares "send energy, light and high-speed particles into space." In essence, there are unimaginably large explosions propelling "solar plasma" which comprises "electrons, protons, and ions" as well as alpha particles out beyond the Sun. It is known in terms of beauty and, for our purposes, nothing more than solar wind.[21]

From the time of the earliest measurements, these winds have tended to occur in 11-year solar cycles.[22] That cycle is also completely skewed in the novel. There is no such cycle, but a continuous series of Z+ CME events occurring day, month, year, and decade after decade. No government has ever adequately prepared for such a "what if?"

For years the impact on earth of solar storms has mostly focused upon the interruption of electrical services: telecom satellites and power grids.[23] Any communication system using high frequency (3–30 MHz) relies on the ionosphere. That comprises a solar ionized shell between 30 and 600 miles in altitude in which TEC (the total electron content) represents a crucial number density through which successful transmission, for example, of Global Positioning Satellites occurs. That TEC is highly vulnerable to solar flares and can result in phase delays of reflected radio signals throughout the world.[24]

The earth's magnetosphere, a vast system that extends from well below the planet's rocky outer core, to a distance "six to ten times the radius of the earth"[25] has shielded the biosphere from excessive radiation but is, at the same time, highly vulnerable to major solar storms. Magnetic poles are constantly shifting and have, in the past totally reversed. Moreover, the magnetosphere continually interacts with the upper atmosphere. A massive CME will alter the electron content in the columns of the magnetosphere and this in turn will affect radio, radar, sonar, GPS, and every other form involving frequencies of communication. It also alters the navigational systems of animals. Migrations of birds and ungulates can be thoroughly skewed, although there is some evidence to suggest that birds can adjust their migratory patterns quite readily in response to environmental changes, as can some ant species.

The common assumption is that solar flares could be dangerous to astronauts, but not to us on earth, because "we're protected by Earth's blanket of atmosphere."[26] This presumption is buttressed by data bases in disciplines including meteorology, atmospheric chemistry, and biochemistry, suggesting that the cycles are normal. However, it is also well attested to that "Very high-energy particles, such as those carried by CME's can cause radiation poisoning to humans and other mammals… Large doses could be fatal."[27] Indeed, almost immediately lethal.[28]

In a world of severe and accelerated climate change, ozone layer depletion, and geomagnetic storms lasting years, even decades, we have no biological baseline in 2022 to even extrapolate the UV penetration levels, and their medical impact on all species. But clearly, melanomas would be widespread, severe, and potentially impactful at species extinction levels. In their migratory confusion, from whales to bees, there would be disastrous consequences in the event of such a scenario.[29] In a worst-case regime, pathogenic loads, "UBV-induced immunosuppression" would likely be very high.[30]

The Melanomas

Aggressive skin cancers, predicating a vast surge of melanoma-related deaths across the entire human population worldwide is part of the CME equation in the novel. The nearly Martian-like preconditions for this to occur on earth has happened repeatedly under very specific conditions, the most serious such event referenced in the book. On December 29, 2003 a UV index of 43.3 occurred atop Bolivia's 19,423-foot stratovolcano, Licancabur. Scientists have noted that this radiation level equated with that on the surface of Mars. Moreover, throughout the Andes, levels in the mid-20's of the Index have routinely occurred. In January 2004, it soared into the 30s. Two weeks prior to those readings, there was a massive solar flare, and many have suggested that it would have contributed to the destruction of the ozone layer had it continued. Putting that in perspective, the WHO (World Health Organization) warns humans to stay indoors if the UV Index should hit 12.[31]

Melanoma incidence is found to be the "most common [of] cancers…in adolescent and young adult populations. In fact, it is one of the leading cancers in average years of life lost per death from disease," write researchers.[32]

The Demographics

As of mid-2021, the human global population was 7.9 billion.[33] Estimates from the United Nations Department of Economic and Social Affairs have recognized that by 2050 humanity should number approximately 9.7 billion, and roughly 11 billion in 2100.[34] Demographic transition has previously been viewed as a rule-of-thumb in human population dynamics: the wealthier a nation becomes, the fewer children it produces. But some have questioned whether there might, in fact, be a demographic transition *reversal* occurring during the twenty-first century.[35] Already we are seeing signs of this taking place in the most populous country in the world, China, which has moved far away from its one-child policy of the 1970s and 80s, to a point now where it is debating incentives for parents to have, if possible, two and three children.[36]

The science of population growth and its impact on the biosphere has been readily summarized by two types of drivers. First, the J-shaped curve produced by exponential growth where per capita population size remains the same, and the population grows faster and faster. Logistic growth, conversely, is represented by an S-shaped curve in which the dominant through-story is the ecological ceiling known as (K), or environmental carrying capacity.[37]

The other crucial factor to understand is that of the I=PAT equation, where I=impact; P=population size; A=the level of affluence of those populations; and T which corresponds to the technological infrastructures and their inflictions on the natural world.[38]

The vast consensus of scientific opinion is that the current population of H. sapiens has already usurped and extracted far more non-renewable resources, either by direct or indirect appropriation, or through destruction, than the earth can sustain. In fact, the WWF-Europe has declared, "If everybody lived as Europeans, we would need 2.8 planets."[39] Such correlations have long been popularized by the environmental community. But there is no single authoritative data set that can actually compute the extent of humanity's onslaught upon the biosphere. A few humans could trigger all-out nuclear war.

Electricity

The engineering, scientific, and technical literature on the collapse of power grids and societal-wide impacts is extensive. That cascade of dysfunctions engenders a predicament confronted by survivors in the novel. What happens when every nation is plunged into a world where there is no electricity? Where satellites, phones, cell tower transmission and relay capacities, every back-up system, are undermined completely? The power at first is erratic. There are ways to induce current, propagate radio frequencies, but not when the entire landscape of electrical generating and transmission infrastructure is eradicated. Over time, batteries are in short supply, and then gone. A radio, and/or Morse Code generator/translator, is also no longer an option, unless there are some vestigial low energy supplies, as indicated in the story.[40]

Excluding all the space junk, there are over 2700 active satellites orbiting earth. If humanity were to lose all of them, in terms of operational reception and transmission of data, at the same time as all power grids on the planet were wiped out, we would simply have to cope with a world comparable to that of a few hundred years ago.[41] We could do that, but not with nearly 8 billion people, let alone 10 or 12 billion. Drinking water, food, and medical assistance would be the three immediate constraints on this scenario, as individuals struggled to survive, in absence of any information outside their immediate neighborhoods. Radio interference would be 100% across every frequency save for the most local ham operations. And even those, in absence of electricity of any kind, or batteries, would go dead.[42]

There is one kink in this otherwise chilling set of circumstances: the radio beacon coming from the museum in Bruges, received at pes (Princess Elisabeth Station) in the Antarctic, as fictionalized in this story. This stems from the suggestion that there was, at least at one time in the near past, a "citizens' satellite" that could be utilized by anyone. Whether it somehow escaped the inferno that took out all the other satellites, is not disclosed in the novel.

Countless simple and complex local technologies have been addressed to help provide for the emergency transmission of information. Discussions of various transmitters, repeaters, foxhole and home-made crystal radios, antennas, ham radio operation, emergency frequencies throughout the world, Skywave, skipping (propagation of radio waves that are transmitted to and from the ionosphere), tropospheric scatter, DXing, line-of-sight propagation, etc. have all been discussed by scientists, engineers, and survivalists in enormous depth. But without electricity, satellites, telephone, and, obviously, the Internet, our species does have a real problem.[43] No other species is all but dependent on technology. Only the poorest of the poor, in terms of current economically marginalized human populations, would stand a chance of weathering a non-technological horizon.

Notwithstanding the many hundreds of personal emergency locator beacons, SOS transmitters, personal satellite communicators, etc.[44] with no satellites and fast waning batteries, production of electricity using hydrogen fuel cell/electrolyzer combinations is a plausible scenario for those duly equipped. But the electricity is restricted. Re-fueling stations are in short supply, and home use would still not get you any information. In reality, a small, confined scientific establishment like an Antarctic station (with a greenhouse and hydroponics) could rely entirely on such technology.[45] The question would be: for how long, and under what atmospheric circumstances? Artificial biospheres (closed ecological life support systems) have not fared well, in the scheme of things; certainly not for any meaningful duration. Add to that imagined bubble the unprecedented escalation of solar emissions, ice shelf collapse, and all the anomalies and biospheric impacts correlated with extreme climate change, and all bets are off.

Future (Human) Society

We see how turbulent and utterly skewed the projections on climate change have been.

Similarly, human demographic data has been rocked repeatedly, with numbers heading north of 10 billion, possibly 11 or 12 billion by the end of the twenty-first century. To date, there has been no psychological or anthropological through-story that has provided a cross-cultural narrative that reasonably explains how to feed so many large, principally carnivorous vertebrates. The vegan way of life is obviously a crucial testament to humanity's ability, albeit slowly, to change its ways: Very slowly (in comparison with the billions of animals being slaughtered daily). The scientific community has worked to improve upon nature's own agricultural mechanisms, including the Marshall Davidson Hatch/Charles Roger Slack findings from the 1960s of four-carbon fixation in photosynthesis among some plants.[46] This has given rise in recent times to countless efforts to invent a second agricultural green revolution. The first one did not turn out so well, as Nobel Laureate Norman Borlaug described in the wake of the first green revolution.[47]

At the same time, the Anthropocene is accelerating in ways, across biomes and species, that were never anticipated. The sixth extinction spasm has come upon us as if, in our blindness to ourselves, it was all of a sudden. When, in reality, it has been building for at least 10,000 years.[48]

All these conflated factors make for a challenging assessment of how human behavior is likely to play out in small hamlets, communities, and megacities of the future.

If we are to survive, will we, in fact, biologically require something akin to the "tameness gene," as indicated in the novel? This would insure (theoretically) a wholly new form of geopolitics and economics that is dependent on local barter systems and outright kindness. An entirely new version of "possessions," socialism, occupancy rights, human and other animal rights in general would come into juridical play (a vast expansion of Habeas corpus briefs on behalf of other individuals of other species, for example; and then an even much further embrace of a fully non-anthropic, non-invasive, Jain Digambara, 'let it be', "live and let live"-like communion). A history of kindness has only recently begun to penetrate the fringes of judicial systems, where the acknowledgment of legal personhood for other species, from great apes to elephants to cephalopods is hotly debated. The fact these debates continue well into the twenty-first century does not bode well at all in terms of projecting whether the ghosts of a Machiavelli and Thomas Hobbes,[49] of a genocidal dictator, are going to remain functionally embedded within some people, some cultures, forever.

If so, given our current technological penchants for destruction (as measured in just the past 120 years), the fixation with Doomsday Clocks, and entertainment promulgated upon scenarios of annihilation, appear coded in our species' unconscious psychological topologies. They, in turn, are undoubtedly part of the genotypic puzzle when it comes to evolution and propensities for change. We embrace fantasies of annihilation, but why? At the same time, we acknowledge loss as a fundamental act of being human.[50]

If ever there were an ecological bubble in which to explore the ecological and sociological stressors of the coming decades, the central marketplace (Markt) in Bruges, Belgium, is certainly a great candidate for assessing human nature, past, present, and future. And our small cast of characters are plausible representatives of the human species grappling to get it right, to figure it out, to survive without further catastrophic inflictions on our only home: earth.

References

1. For the most recent ice shelf collapse concerns, see "One of Antarctica's Largest Ice Shelves Is About To Collapse, New Study Says," by Carlie Porterfield, April 28, 2021, Forbes, https://www.forbes.com/sites/carlieporterfield/2021/04/28/one-of-antarcticas-largest-ice-shelves-is-about-to-collapse-new-study-says/?sh=2d347c6e31a5; See also, "Four decades of Antarctic Ice Sheet mass balance from 1979–2017," by Eric Rignot, Jérémie Mouginot, Bernd Scheuchl, Michiel van den Broeke, Melchior J. van Wessem, and Mathieu Morlighem, PNAS January 22, 2019 116 (4) 1095–1103; first published January 14, 2019; https://doi.org/10.1073/pnas.1812883116, Proceedings of the National Academy of Sciences of the United States of America, https://www.pnas.org/content/116/4/1095. Shockingly, countless new species are recorded in various heretofore unexplored regions of Antarctica just as they probably go extinct. The habitat in sub-freezing waters beneath Antarctica represents many such areas. Writes Isaac Schultz, "So far, only about 10 square feet (1 square meter) of the 620,000-square-mile (1.6-million-square-kilometer) habitat has actually been observed, leading to fears that some of the biodiversity under Antarctica could undergo anonymous extinction." See "Scientists Found a Cradle of Life Under Antarctica," by Isaac Schultz, December 21, 2021, Gizmodo, https://gizmodo.com/scientists-found-a-cradle-of-life-under-antarctica-1848252604/amp

2. "Antarctic Ocean Reveals New Signs of Rapid Melt of Ancient Ice, Clues About Future Sea Level Rise," by Bob Berwyn, May 28, 2020, Inside Climate News, https://insideclimatenews.org/news/28052020/antarctic-ocean-ice-melt-climate-change/

3. See "Crucial Antarctic ice shelf could fail within five years, scientists say," The Washington Post, Sarah Kaplan, December 13, 2021, https://www.washington-post.com/climate-environment/2021/12/13/thwaites-glacier-melt-antarctica/ See also, "Future climate response to Antarctic Ice Sheet melt caused by anthropogenic warming," by Shaina Sadai, Alan Condron, Robert DeConto, and David Pollard, Science Advances 23 Sep 2020:Vol. 6, no. 39, eaaz1169, DOI: https://doi.org/10.1126/sciadv.aaz1169, https://advances.sciencemag.org/content/6/39/eaaz1169; See also, "'Uncertainty is not our friend': Scientists are still struggling to understand the sea level risks posed by Antarctica," by Chris Mooney and Brady Dennis, The Washington Post, May 5, 2021,https://www.washingtonpost.com/climate-environment/2021/05/05/uncertainty-is-not-our-friend-scientists-are-still-struggling-understand-sea-level-risks-posed-by--antarctica/; See also, "Antarctic ice loss has tripled in a decade. If that continues, we are in serious trouble." By Chris Mooney, June 13, 2018, The Washington Post, https://www.washingtonpost.com/news/energy-environment/wp/2018/06/13/antarctic-ice-loss-has-tripled-in-a-decade-if-that-continues-we-are-in-serious-trouble/

4. "Greenland and Antarctica Already Melting at 'Worst-Case-Scenario' Rates," by Olivia Rosane, September 3, 2020, EcoWatch, https://www.ecowatch.com/greenland-antarctica-ice-melting-rate-2647425507.html; See also, "Global warming pushing emperor penguins to brink of extinction by 2100," by Miki Perkis, The Sydney Morning Herald, August 5, 2021,https://www.smh.com.au/environment/climate-change/global-warming-pushing-emperor-penguins-to-brink-of-extinction-by-2100-20210805-p58g01.html; Of the sixty remaining colonies of Emperors on the continent, all are listed by the IUCN Red List as "Threatened". See also, "Every penguin, ranked: which species are we most at risk of losing?" by Alex Dale, BirdLife International, April 24, 2017,https://www.birdlife.org/list-penguin-species; See also, Red List, "Emperor Penguins and Climate Change,"https://www.iucn.org/sites/dev/files/import/downloads/fact_sheet_red_list_emperor_v2.pdf; See also, "Antarctica's 'Doomsday Glacier' is fighting an invisible battle against the inner Earth, new study finds," by Brandon Specktor, LiveScience, April 14, 2021,https://www.livescience.com/doomsday-glacier-close-to-tipping-point.html; See also, "Rain fell at the normally snowy summit of Greenland for the first time on record," by Rachel Ramirez, CNN, August 19, 2021, https://www.cnn.com/2021/08/19/weather/greenland-summit-rain-climate-change/index.html

5. "Fauna of the Antarctic," Umwelt Bundesamt, January 27, 2016, https://www.umweltbundesamt.de/en/fauna-of-the-antarctic#krill-as-the-basis-for-the-food-chain; See also, "Surviving out in the Cold: Antarctic Endemic Invertebrates and Their Refugia," by Malte Ebach, P.J.A.Pugh, and P. Convey, Journal of Biogeography, Vol. 35, No. 12 (Dec., 2008), pp. 2176–2186, Wiley

Publishers, https://www.jstor.org/stable/20488312, https://www.jstor.org/stable/20488312; See also, "Antarctica," Living National Treasures, http://lntreasures.com/antarctica.html; See also, "Plants And Microbes," https://www.antarctica.gov.au/about-antarctica/plants/

6. "Princess Elisabeth Antarctica," http://www.antarcticstation.org/

7. See Cobb RE, Chao R, Zhao H (May 2013). "Directed Evolution: Past, Present and Future". AIChE Journal. 59 (5): 1432–1440. doi: https://doi.org/https://doi.org/10.1002/aic.13995. PMC 4344831. PMID 25733775. See also, Romero PA, Arnold FH (December 2009). "Exploring protein fitness landscapes by directed evolution". Nature Reviews. Molecular Cell Biology. 10 (12): 866–76. doi: https://doi.org/https://doi.org/10.1038/nrm2805. PMC 2997618. PMID 19935669.

8. See "'Test-tube' evolution wins Chemistry Nobel Prize," by Elizabeth Gibney, Richard Van Noorden, Heidi Ledford, Davide Castelvecchi & Matthew Warren, Nature, News, 03 October 2018, https://www.nature.com/articles/d41586-018-06753-y

9. "Meet the Scientists Bringing Extinct Species Back From the Dead," by Amy Dockser Marcus, October 11, 2018, The Wall Street Journal, https://www.wsj.com/amp/articles/meet-the-scientists-bringing-extinct-species-back-from-the--dead-1539093600; See also,https://reviverestore.org/advances-in-avian--transgenics-a-follow-up-to-why-birds-are-a-challenge/; andhttps://www.scientificamerican.com/article/scientists-have-reconstructed-the-genome-of-a--bird-extinct-for-700-years/little_bush_moa; andhttps://www.npr.org/sections/thesalt/2019/01/03/681941779/scientists-have-hacked-photosynthesis-in-search-of-more-productive-crops

10. "What is Rewilding?" https://truenaturefoundation.org/what-is-rewilding/; See also, "RewildingEarth", https://rewilding.org/what-is-rewilding/; See also, https://www.half-earthproject.org/

11. See "6 extinct animals that could be brought back to life," by Live Science Staff, March 15, 2013, https://www.livescience.com/27930-images--deextinction-species.html; See also, "Japanese scientists make breakthrough in cloning a woolly mammoth," https://www.dw.com/en/japanese-scientists--make-breakthrough-in-cloning-a-woolly-mammoth/a-48063060; See also, the important ethical implications of de-extinction, as described in David Shultz' essay, "Should we bring extinct species back from the dead?" Science, September 26, 2016, https://www.sciencemag.org/news/2016/09/should-we-bring-extinct-species-back-dead

12. For this important study, see Kukekova, A.V., Johnson, J.L., Xiang, X. et al. Red fox genome assembly identifies genomic regions associated with tame and aggressive behaviours. Nat Ecol Evol 2, 1479–1491 (2018). https://doi.org/10.1038/s41559-018-0611-6, September 2018, https://doi.org/10.1038/s41559-018-0611-6, https://www.nature.com/articles/s41559-018-0611-6#citeas

13. G.M.Vaquero, M. Vazquez, (2009) The Sun Recorded Through History. Springer New York.

14. Usoskin, I. (2017). "A history of solar activity over millennia". Living Reviews in Solar Physics. 14 (1): 3. arXiv:0810.3972.Bibcode:2017LRSP...14....3U. See also, Wu, C.-J.; et al. (2018). "Solar total and spectral irradiance reconstruction over the last 9000 years". Astron. Astrophys. 620: A120. arXiv:1811.03464. Bibcode:2018A&A...620A.120W.doi:https://doi.org/https://doi.org/10.1051/0004-6361/201832956.S2CID 118843780. See also, Petrovay, K. (2019). "Solar cycle prediction". Living Reviews in Solar Physics. 7: 6.arXiv:1012.5513. Bibcode:2019arXiv190702107P. doi: https://doi.org/https://doi.org/10.12942/lrsp-2010-6.PMC 4841181.PMID 27194963.

15. Boteler, D. H., Pirjola, R. J. and Nevanlinna, H., The effects of geomagnetic disturbances on electrical systems at the Earth's surface. Adv. Space. Res., 22(1), 17-27, 1998. See also, OECD/IFP Futures Project on "Future Global Shocks,"—Geomagnetic Storms, CENTRA Technology, Inc., on behalf of Office of Risk Management and Analysis, U.S. Dept. of Homeland Security, https://www.oecd.org/governance/risk/46891645.pdf

16. Pirjola, R., Fundamentals about the flow of geomagnetically induced currents in a power system applicable to estimating space weather risks and designing remedies. J. Atmos. Sol. Terr. Phys., 64(18), 1967–1972, 2002.

17. See Bolduc, L., GIC observations and studies in the Hydro-Québec power system. J. Atmos. Sol. Terr. Phys., 64(16), 1793–1802, 2002. [Google Scholar] See also, "A 100-year solar storm could fry our power grids—these are the places most at risk," by Dave Mosher and Andy Kiersz, September 19, 2016, INSIDE, https://www.businessinsider.com/solar-storm-risk-map-united-states-2016-9 "What would happen in an apocalyptic blackout?" Richard Gray, BBC, October 24, 2019, https://www.bbc.com/future/article/20191023-what-would-happen-in-an-apocalyptic-blackout

18. See "How We'll Safeguard Earth From a Solar Storm Catastrophe," by Rebecca Boyle, June 8, 2017, https://www.nbcnews.com/mach/space/how-we-ll--safeguard-earth-solar-storm-catastrophe-n760021, NBC, MACH; See also, Carrington, R. C. (1859). "Description of a Singular Appearance seen in the Sun on September 1, 1859". Monthly Notices of the Royal Astronomical Society. 20: 13–5. Bibcode:1859MNRAS..20...13C. doi: https://doi.org/https://doi.org/10.1093/mnras/20.1.13.

19. https://www.nasa.gov/mission_pages/sunearth/news/X-class-flares.html

20. "What Damage Could Be Caused by a Massive Solar Storm?" by Joseph Stromberg, Smithsonian Magazine, February 22, 2013, https://www.smithsonianmag.com/science-nature/what-damage-could-be-caused-by-a-massive-solar-storm-25627394/

21. "What Does It Take To Be X-Class?" NASA, August 9, 2011, https://www.nasa.gov/mission_pages/sunearth/news/X-class-flares.html

22. op cit., J.Strombert, 2013.

23. B. Hofmann-Wellenhof; H. Lichtenegger & J. Collins (2001). Global Positioning System: Theory and Practice. New York: Springer-Verlag.

24. "Magnetospheres," NASA Science, SUN, https://science.nasa.gov/heliophys-ics/focus-areas/magnetosphere-ionosphere/; See also,https://www.livescience.com/amp/solar-storm-internet-apocalypse

25. See "Are Solar Storms Dangerous To Us?" by Deborah Byrd, EarthSky, January 30, 2020, https://earthsky.org/space/are-solar-storms-dangerous-to-us/; Regarding polar destabilization, see "Reversal of Earth's magnetic poles may have triggered Neanderthal extinction—and it could happen again," Amy Woodyatt, CNN, February 19, 2021,https://www.cnn.com/2021/02/19/world/magnetic-fields-earth-intl-scli-scn/index.html; See also, "Climate change has destabilized the Earth's poles, putting the rest of the planet in peril, The Washington Post, by Sarah Kaplan, December 14, 2021,https://www.washing-tonpost.com/climate-environment/2021/12/14/climate-change-arctic-antarctic-poles/; See also, The National Oceanic and Atmospheric Administration (NOAA) released 2021's Arctic report card on Dec. 14. (NOAA); See also, "The Arctic could get more rain and less snow sooner than projected. Here's why that matters." Brady Dennis and Kasha Patel, Washington Post, November 30, 2021, https://www.washingtonpost.com/climate-environment/2021/11/30/arctic-rain-snow-climate-change/

26. ibid.

27. Council, National Research; Sciences, Division on Engineering and Physical; Board, Space Studies; Applications, Commission on Physical Sciences, Mathematics, and Research, Committee on Solar and Space Physics and Committee on Solar-Terrestrial (2000). Radiation and the International Space Station: Recommendations to Reduce Risk. National Academies Press. p. 9.

28. https://www.usnews.com/news/national-news/articles/2017-09-06/solar-storms-may-ignite-south-reaching-auroras-wednesday

29. See "Risk assessment for the harmful effects of UVB radiation on the immuno-logical resistance to infectious diseases," by W Goettsch, J Garssen, W Slob, F R de Gruijl, and H Van Loveren, Environmental Health Perspectives, Environ Health Perspect. 1998 Feb; 106(2): 71–77. doi: https://doi.org/https://doi.org/10.1289/ehp.9810671, PMCID: PMC1533030, PMID: 9435148, https://www.ncbi.nlm.nih.gov/pmc/articles/PMC1533030/

30. "Record solar UV irradiance in the tropical Andes," by Nathalie A. Cabrol, Uwe Feister, Donat-Peter Hader, Helmut Piazena, Edmond A. Grin and Andreas Klein, Frontiers in Environmental Science, 08 July 2014 https://doi.org/10.3389/fenvs.2014.00019, https://www.frontiersin.org/articles/10.3389/fenvs.2014.00019/full; See also, "Blazing world record: Strongest UV rays ever mea-sured om Earth", by Becky Oskin, LiveScience.com, July 8, 2014, https://www.cbsnews.com/news/blazing-world-record-strongest-uv-rays-measured-in-south-america/

31. "Epidemiology of melanoma," by Natalie H. Matthews, Wen-Qing Li, Abrar A. Qureshi, Martin A. Weinstock, and Eunyoung Cho, Chapter One, in Cutaneous Melanoma—Etiology and Therapy, Edited by William H. Ward and Jeffrey M. Farma, Codon Publications, December 21, 2017, Bribane, AU, https://www.ncbi.nlm.nih.gov/books/NBK481860/ See also, "How UV

Radiation Triggers Melanoma," from NIH Research Matters, February 7, 2011, https://www.nih.gov/news-events/nih-research-matters/how-uv-radiation-triggers-melanoma

32. "World Population Clock," https://www.worldometers.info/world-population/
33. https://www.un.org/development/desa/en/news/population/world-population--prospects-2019.html
34. Can we be sure the world's population will stop rising?, BBC News, 13 October 2012.
35. "China allows three children in major policy shift," Analysis by Stephen McDonell, May 31, 2021, BBC News, https://www.bbc.com/news/world-asia-china-57303592
36. See "Exponential & Logistic growth," Khan Academy, https://www.khanacademy.org/science/ap-biology/ecology-ap/population-ecology-ap/a/exponential--logistic-growth. See also, "How Many People Can Earth Support?" by Natalie Wolchover, LiveScience, October 11, 2011, https://www.livescience.com/16493-people-planet-earth-support.html; See also the latest demographic health surveys, as well as population projections at, http://populationcommunication.com/
37. See "A brief history of 'IPAT' (impact = population x affluence x technology)", by John P. Holdren, The Journal of Population and Sustainability, Vol 2, No 2, 2018, https://jpopsus.org/full_articles/holdren-vol2-no2/
38. EcoWatch, https://www.ecowatch.com/eu-overshoot-day-2636803420.html; See also, "Earth Overshoot Day: We Consumed 12 Months Worth of the Earth's Resources in Just 7 Months", by Jennifer Nini, ecowarrior, February 21, 2018.
39. See "Huge Solar Flare's Magnetic Storm May Disrupt Satellites, Power Grids," by Denise Chow, March 7, 2012, https://www.space.com/14818-solar-flare--magnetic-storm-satellites.html
40. "What Would Happen If All Our Satellites Were Suddenly Destroyed?" by George Dvorsky, Gizmodo, June 4, 2015, https://gizmodo.com/what-would--happen-if-all-our-satellites-were-suddenly-d-1709006681; See also, https://www.pilotfiber.com/blog/antarctica-internet
41. See "Identifying and Locating Radio Frequency Interference (RFI)", September 7, 2018, by Kenneth Wyatt, Interference Technology, https://interferencetechnology.com/author/kennethwyatt/
42. See "How To Communicate When the World Goes Silent," graywolfsurvival.com, https://graywolfsurvival.com/2716/ham-radio-best-shtfdisaster-communication/; See also, https://www.offthegridnews.com/extreme-survival/4-life-saving-ways-to-communicate-when-the-power-is-out/; See also, Seybold, John S. (2005). Introduction to RF Propagation. John Wiley and Sons. pp. 3–10. See also, "The Electrification of the World—Werner von Siemens and the Dynamoelectric Principle," Siemens Historical Institute, https://new.siemens.com/global/en/company/about/history/stories/dynamo-machine.html
43. "How Humanity Came To Contemplate Its Possible Extinction: A Timeline," by Thomas Moynihan, September 23, 2020, https://thereader.mitpress.mit.edu/how-humanity-discovered-its-possible-extinction-timeline/

44. See https://www.amazon.com/personal-locator-beacon/s?k=personal+locator+beacon
45. See "Hydrogen Explained," eia, U.S. Energy Information Administration, https://www.eia.gov/energyexplained/hydrogen/use-of-hydrogen.php
46. Slack CR, Hatch MD (June 1967). "Comparative studies on the activity of carboxylases and other enzymes in relation to the new pathway of photosynthetic carbon dioxide fixation in tropical grasses". The Biochemical Journal. 103 (3): 660–5. doi:https://doi.org/https://doi.org/10.1042/bj1030660. PMC 1270465. PMID 4292834. See also, "The Control Of Nature—A New Leaf—Could tinkering with photosynthesis prevent a global food crisis?" by Elizabeth Kolbert, The New Yorker, December 13, 2021, pp. 30–36.
47. See "The Green Revolution: Norman Borlaug and the Race to Fight Global Hunger," by Ray Offenheiser, The American Experience, PBS, April 3, 2020, https://www.pbs.org/wgbh/americanexperience/features/green-revolution-norman-borlaug-race-to-fight-global-hunger/
48. See *Anthrozoology: Embracing Co-Existence in the Anthropocene*, by M. C. Tobias and J. G. Morrison, Springer Nature, New York, NY, 2017.
49. See *Leviathan or The Matter, Forme and Power of a Commonwealth Ecclesiasticall and Civil*, by Thomas Hobbes (1588–1679) published in 1651 (London, and Amsterdam).
50. See "Australian Bird Calls—Songs of Disappearance," Album by David Stewart and The Bowerbird Collective, December 2, 2021, https://music.amazon.com/albums/B09LYBN4TQ; See also, *The Annihilation of Nature—Human Extinction of Birds and Mammals*, by Gerardo Ceballos, Anne H. Ehrlich, and Paul R. Ehrlich, Johns Hopkins University Press, Baltimore, MD, 2015.

Printed in the United States
by Baker & Taylor Publisher Services